MIKE LOW recently left Rolls a working lifetime involved in engines and other power plants Nuclear sectors. It was from this that Mike became interested in th............... effect on the outcome of World War II made by the Merlin engine.

Mike has recently published *The Bleeding Obvious Way to Improve Quality in Your Business,* a book about the basics of good quality practice. He is married and lives in Somerset.

To find out more about the author, visit his website at: www.mikelow.co.uk.

THE STORY OF THE
ROLLS-ROYCE MERLIN AERO ENGINE

SAVIOUR
OF THE
FREE
WORLD

MIKE LOW

SilverWood

Published in 2015 by SilverWood Books

SilverWood Books Ltd
14 Small Street, Bristol, BS1 1DE, United Kingdom
www.silverwoodbooks.co.uk

Copyright © Mike Low 2015

Interior images © Rolls Royce Heritage Trust Publications
(p.45–7; p.142; p.158: top), Getty Images (p.117: top), FlightGlobal (p.117: middle), National Portrait Gallery (p.118: bottom right; p.157: top), Diomedia (p.118: top right), Imperial War Museum (p.156: bottom), Wikimedia Commons (p.116: middle; p.158: bottom), Royal Canadian Air Force Bomber Command Museum (p.119: bottom).

The author has made reasonable efforts to trace the current holders of the copyright of material used in this book. If any copyright holder believes that their material has been used without due credit, the author will make reasonable efforts to correct omissions in any future re-printing.

The right of Mike Low to be identified as the author of this work has been asserted by him in accordance with the Copyright, Designs and Patents Act 1988.

All rights reserved. No part of this publication may be reproduced, stored in a retrieval system, or transmitted in any form or by any means, electronic, mechanical, photocopying, recording or otherwise, without prior permission of the copyright holder.

ISBN 978-1-78132-395-3

British Library Cataloguing in Publication Data
A CIP catalogue record for this book is available from the British Library

Set in Sabon by SilverWood Books
Printed on responsibly sourced paper

*To my wife, for her constant support at all times.
Thank you, darling.*

To be ignorant of what occurred before you were born is to remain always a child.

Marcus Tullius Cicero

I expect that the Battle of Britain is about to begin. Upon this battle depends the survival of Christian civilisation. Upon it depends our own British life, and the long continuity of our institutions and our Empire. The whole fury and might of the enemy must very soon be turned on us. Hitler knows that he will have to break us in this island or lose the war. If we can stand up to him, all Europe may be free and the life of the world may move forward into broad, sunlit uplands.

But if we fail, then the whole world, including the United States, including all that we have known and cared for, will sink into the abyss of a new dark age made more sinister, and perhaps more protracted, by the lights of perverted science. Let us therefore brace ourselves to our duties, and so bear ourselves, that if the British Empire and its Commonwealth last for a thousand years, men will still say, this was their finest hour.

Winston Churchill
Speech to the House of Commons
18 June 1940

Contents

Acknowledgements	12
Foreword	15
Chapter 1: A Chequered Provenance	19
Eagle Engine	
Kestrel Engine	
R Engine	
Schneider Trophy	
PV12 Engine	
Chapter 2: The Merlin	34
Background History	
The Supercharger	
The Cooling System	
One Hundred Octane Fuel	
Failure Investigation	
Development History	
The Difference	
A Perspective	
Chapter 3: The Competition	53
Friendly	
Unfriendly	
Chapter 4: Work Until it Hurts	59
The Factories That Produced the Merlin Engine	
Rolls-Royce Derby	
Rolls-Royce Glasgow Hillington	
Rolls-Royce Crewe	
Ford Trafford Park, Manchester	
Packard Motor Car Company of Detroit	

Chapter 5: Inspirational Leadership ... 71
    Ernest Hives
    Wilfrid Freeman and Lord Beaverbrook

Chapter 6: The Architects of Victory ... 79
    Lucy, Lady Houston
    Henry Royce
    Charles Rolls
    Cyril Lovesey
    Arthur Rubbra
    Stanley Hooker
    Beatrice Shilling

Chapter 7: Aircraft Designers Who Used the Merlin ... 103
    Reginald Mitchell
    Roy Chadwick
    Sidney Camm
    Ronald Bishop
    Edgar Schmued

Chapter 8: Aircraft ... 120
    Spitfire and Hurricane
    Lancaster
    Mosquito
    P-51 Mustang

Chapter 9: Pilots ... 132
    George Stainforth
    Augustus Orlebar
    Guy Gibson
    Douglas Bader
    Ronald Harker
    The Memorial Window

Chapter 10: More Than an Aero Engine and Beyond ... 145
    Meteor Tank Engine
    Marine Merlin
    Beginnings of Gas Turbine Power

Chapter 11: Today	152
Merlin Repair and Overhaul	
Battle of Britain Memorial Flight	
P-51 Mustang	
Chapter 12: What If?	160
Hindsight	
Hitler's Directive Number 16	
Possible Outcomes	
Alternative Assessments	
Other Factors	
Chapter 13: What Now?	185
Engineering in Britain Today	
Last Word	
Postscript	191
Merlin (*Falco columbarius*)	
Addendum A	192
Merlin Powered Aircraft	
Addendum B	194
Merlin Engine Cutaway Diagram	
Addendum C	195
Top 20 Engineering Heroes	
Glossary	198
Notes on Engineering Terms	
Conversion Tables	203
Bibliography	204
Endnotes	210

Acknowledgements

The inspirations for this book were many and varied. They included my years working for one of the greatest engineering businesses in the world, Rolls-Royce, and the support of the many friends and associates with whom I was fortunate enough to collaborate during my years working there. One of the other main inspirations for this account was the book *Engineers of Victory* by the historian Paul Kennedy. It is a stimulating history explaining the vital contribution of problem solvers and engineers to the successful outcome of World War II.

Numerous sources were consulted during research for this history. I have endeavoured to ensure that every fact and quote from those references is verified by documentary or Internet evidence from at least two separate sources in order to ensure historical accuracy. Any errors or omissions are my own.

My thanks go to Tony Griffin for contributing many relevant and interesting books from his comprehensive library; to Jim Napier and Arthur Spencer for the tales of their experiences when sitting behind four Merlins; to

Adrian Hart at SilverWood Books for his useful suggestions and to Jacqueline Morris for her help in compiling the photographs.

The series of historical and technical books from the Rolls-Royce Heritage Trust[1] has also been of great help in compiling this history. Anyone interested in finding out more about some of the characters and events around the production and development of the great Merlin engine would do well to contact the Trust for more information on their range of books. Their address is given in the endnotes.

Note: Imperial measurements are used throughout this volume. There is a conversion table to metric equivalents at the back of the book.

Foreword

This book tells the story of the Rolls-Royce Merlin aero engine, from its earliest beginnings through development and deployment in the aircraft that fought the Battle of Britain and a great variety of other aircraft that secured victory and freedom during World War II, and onward to today in the Battle of Britain Memorial Flight. It is a story of vision, determination, persistence, innovation and triumph. There can be little doubt that without the Merlin engine, the Battle of Britain may well have been lost and the islands of Britain could have been overrun by the Nazis.

This is also the story of those engineers, production workers, managers and others who 'Worked Until it Hurt'[2] to produce the world-beating Merlin engine in such prodigious quantities and contributed to its success; the roles of Sir Henry Royce, Lord Hives and some of Britain's greatest-ever engineers: Sir Stanley Hooker, A.A. Rubbra, S. Camm, R.J. Mitchell and Beatrice Shilling, without whom our island history, and the history of the free world, might have been very different.

One could not tell the story of this outstanding engine without briefly touching upon the immense contribution of the pilots of the Royal Air Force and other air forces, who by their bravery and resolve succeeded in realising the full potential of the Merlin: men like Ronnie Harker, Douglas Bader and Guy Gibson. A short biography of each of these great men is included. Although the other RAF pilots in the Battle of Britain and throughout the war deserve an account, for reasons of space we focus on just a few.

A description of Major G.P. Bulman's involvement in the development of the Merlin is included. Bulman is a relatively unknown and unsung hero, who may well have helped to decide the fate of the free world in a few short conversations by directing a design genius down the right path. Also included is an account of the life of Lucy, Lady Huston, without whose patriotism and generosity – well, I am dismayed at the thought.

The histories of the Merlin and the Rolls-Royce Company are inseparably linked. Without the Merlin, Rolls-Royce would not be the business it is today. The Merlin made an unrivalled contribution to the Allied cause in World War II, and for Rolls-Royce it caused a metamorphosis. Having entered the war as a medium-sized high quality manufacturer in a Derby side road, it emerged as a major British company, and today it is a foremost global engineering organisation.

All of us are shaped by our experiences. My first appreciation of the importance of the Merlin engine, though I did not realise it at that time, (and one of my earliest memories of moving pictures as my family did not have a TV then) was being taken by my father as a seven-year-old to the Regal cinema in Bristol to watch *The Dam Busters*. This film tells the story of the World War II bombing raid by RAF 617 Squadron on the Möhne, Eder and Sorpe Dams in Germany in 1943 using Barnes Wallis's 'bouncing bomb'. I have never forgotten what was, for me, a brand new experience, and whenever the film is shown on TV it still has an intense effect. At that time my horizon was limited to the experience of being in a 'grown-up' cinema, and seeing the bravery and sacrifice of the 617 Squadron pilots, the success of their operation, and the ultimate sacrifice made by the crews of the eight Lancaster bombers lost during the raid. Much later, having become an engineer working for Rolls-Royce, I was left pondering the importance of the engines that powered the Lancaster bombers used on that raid. I began to realise and appreciate the importance of the Merlin engine to the Allied war effort in the Battle of Britain, and in every sphere in which it was used during World War II.

This book is a grouping of many histories and mini-biographies, some of which may be familiar to the reader and others unfamiliar. It is a combination of the history of the Merlin and the history of the people involved in its development, production and realisation through its use in the many aircraft it powered. I am as aware as it is possible to be of the momentous work done by the engineers, managers and workers of every description at that dark time. We are indebted to them all.

Chapter 1
A Chequered Provenance

This book shows how Hitler's plans for world domination were frustrated primarily by the deployment of the most effective engine used throughout World War II, and particularly in the Battle of Britain; the machine that powered more aircraft than any other during the war; the engine that was the mainstay of Royal Air Force Bomber Command and Fighter Command: the Merlin.

The history begins in the early years of the Twentieth Century.

Eagle Engine

At the outbreak of World War I in August 1914, the Royal Aircraft Factory, a government agency, (today known as the Royal Aircraft Establishment, based at Farnborough) asked Rolls-Royce to develop a new 200hp air cooled engine to meet the expected demand of the war. This was agreed with the Royal Aircraft Factory, but only on condition that the engine be liquid cooled. The Rolls-Royce engineers were more familiar with liquid cooling of engines as used in their automotive power plants. The

Admiralty ordered twenty-five of the engines in January 1915, and the Eagle became the first aero engine to be developed by Rolls-Royce. So began the long involvement of Rolls-Royce in supplying aero engines to the Royal Air Force that continues to this day.

Progress in the design of the new engine was led by Henry Royce, and was based initially on the 1908 Rolls-Royce Silver Ghost (originally named the 40/50hp) six-cylinder car engine that developed 50 horse power (hp). The necessary power increase required by an aero engine was achieved by doubling the number of cylinders to twelve, arranging them in a V configuration and increasing the cylinder stroke to 6.5inches. These changes eventually allowed the Eagle to develop 360hp. The first application of this engine was in the Handley Page Type O bomber, a twin-engine biplane, followed by use in a number of other aircraft including the Supermarine Sea-Eagle, a civilian passenger carrying transport plane, and the Handley Page V/1500, a night flying four engine heavy bomber built towards the end of World War I. The Allies intended to use the V/1500 to bomb Berlin, but the end of hostilities prevented its use against Germany. Production of the engine continued until 1928, and altogether 4,681 Eagle engines were built.

Famously, in June 1919 the Eagle engine powered the twin-engine Vickers Vimy bomber, piloted by John Alcock and Arthur Brown, on the first direct flight across the Atlantic. Their crossing from Newfoundland to County Galway in Ireland took sixteen hours and twenty-seven minutes.

A smaller version of the Eagle, the Rolls-Royce Falcon engine, was developed in 1916 and continued in production until 1927, by which time a total of 2,185 had been built. It produced 288hp and powered other World War I aircraft, including the Vickers Viking, a single-engine amphibious aircraft designed and developed for military use shortly after World War I, and the Bristol F2 fighter, which was used by over thirty Royal Flying Corps Squadrons during World War I. (The RFC was renamed the Royal Air Force in April 1918.) The Falcon was used extensively throughout the world until well into the 1930s. Over 5,000 Bristol F2s were built, however, many had to be powered by Hispano-Suiza and Sunbeam engines because there was a shortage of supply of Falcon engines, due mainly to the demands of war.

The Rolls-Royce Company found itself in a less than ideal position in the 1920s in regard to leadership in aero engine development, and a new, more powerful engine was needed to reassert the company's reputation. The Eagle engine had proved its capabilities during the conflict. However, as the world moved into the 1920s, it was not powerful enough to compete with other foreign-engine aircraft at the very highest level in terms of power output and speed generated. A more powerful engine was now required if Britain was to compete with the rest of the world, particularly in the Schneider Trophy air races.

Kestrel Engine

In the early- and mid-1920s, following a deliberate policy of almost abandoning aero engine development in order to focus effort on automobile manufacture, Rolls-Royce was no longer the leader in the aero engine field. They had slipped to third place behind the Bristol Engine Company and the D. Napier and Sons Company. The development of the Kestrel engine was important in re-establishing Rolls-Royce as the foremost supplier of engines to the RAF. Designed by Arthur Rowledge, the Kestrel followed the Rolls-Royce tradition for aero engines in being of a V-12 configuration, which used supercharging to improve performance at all altitudes, allowing it to outperform other engines by increasing the boost pressure.

The engine was first produced in 1927 at 450hp and saw widespread use in the Hawker Hart, a two-seater light bomber of the RAF that was the mainstay of British air power during the early 1930s. The Kestrel also powered the Miles Master two-seater monoplane advanced trainer. The engine was modified and developed, eventually producing 700hp and achieving a significant reputation in the history of aviation. There were 4,750 engines produced before production ceased.

By the outbreak of World War II, to a large degree many of Britain's RAF pilots owed their fitness in the ensuing struggle to the fact that they had been trained on aircraft powered by the Kestrel, including the Hawker Hart, Hawker Fury and Fairey Firefly. These biplanes were all to be replaced by Merlin-engine Hurricanes and Spitfires before the conflict began. The Kestrel continued to perform an important role throughout the war, however, as it was the engine in the Miles Master advanced trainer used for training RAF pilots.

Remarkably, in 1935 The RLM (*Reichsluftfahrtministerium* – Reich Air Ministry of Germany) acquired four Kestrel engines from Rolls-Royce by trading in exchange a Heinkel He70 aircraft. The Heinkel was to be used by Rolls-Royce as an engine test bed and by the Supermarine Company as a model of how to improve the aerodynamic smoothness of its aircraft. The traded Kestrel engines were used by Messerschmitt to power the first Bf109 fighter prototype because the German engine that had been intended for use was not ready. Similarly, the Junkers Ju87 Stuka dive-bomber also used a Kestrel engine on its first development prototype. The Heinkel was used by Rolls-Royce to test the Merlin's immediate predecessor, the PV12. Had both sides only known what was to come, they might not have been quite so accommodating to each other!

The Kestrel engine did reassert the reputation of Rolls-Royce as a premier aero engine manufacturer. What was now needed as the 1920s drew to a close was an even more special engine that would be able to compete in air racing at the highest level.

R Engine

As the 1920s moved on, it became apparent to the Supermarine Company and their engineers that a more powerful engine was required to contend against the French, Americans and Italians in the Schneider Trophy air races. Accordingly, requests were made to Rolls-Royce to develop and build a new racing engine. These appeals came from Major Bulman of the Air Ministry and R.J. Mitchell, the Chief Designer at Supermarine.

In October 1928, Henry Royce, at his home in West Wittering, Sussex, met with two of his leading engine designers, A.J. Rowledge and A.C. Lovesey, and the Head of Rolls-Royce Experimental Department, E.W. Hives.

All three men were in favour of developing an engine to power the Supermarine seaplanes in the quest to win the Schneider Trophy series of international seaplane races. Royce himself was enthusiastic, and he suggested a stroll along the nearby beach to discuss the project. Royce was by then a semi-invalid and quickly became tired, so the group stopped to rest and talk about engines. Royce sketched a rough outline of the proposed engine in the sand with a stick, and each man gave his opinion in turn. As each view was discussed, the sand would be raked over and adjustments made to the proposed design. The basics were decided upon and, like the Kestrel engine which was first produced in 1927, the new engine would have only twelve cylinders. The bore and stroke would be 6 inches by 6.6 inches, and the compression ratio six to one. The key to the new engine would be simplicity.

"I invent nothing. Inventors go broke"[3] was Sir Henry Royce's philosophy. The secret of increasing the power of the engine would lie in supercharging. The engine was to be known as the R engine ('R' stood for racing).

The R engine was subsequently designed by a team led by Arthur Rowledge of Rolls-Royce. Rowledge was the Chief Designer at aero engine manufacturer Napier and Sons in 1913 and was responsible for designing the Napier Lion engine. In 1921 Rowledge became Chief Designer with Rolls-Royce. One of his last jobs at Rolls-Royce involved development work on the Merlin. Rowledge retired in 1945, and died in 1957.

Built specifically for air racing in the Supermarine seaplanes for the Schneider Trophy races, only nineteen R engines were constructed between 1929 and 1931. The first R engine ran on 7 April 1929. It was a 37 litre capacity supercharged V12 engine that could produce in excess of 2,500hp. At that time, 2,500hp from one engine was an undreamed of possibility to most engineers. The impetus for its design and production was based on the desire, by all concerned with its manufacture, to win the Schneider Trophy outright for Great Britain.

The total commitment of everyone involved in the development of this engine extended to those who were not directly involved at all – the general population of Derby. Throughout the summer of 1929 they endured the testing of the R engine day and night. The open unsilenced exhausts produced a noise that could be heard over large parts of Derby, and became known as the 'Derby Hum'. Testing also required the simultaneous running of three Kestrel engines, which were used to drive the large fans that cooled the R engine and remove the exhaust fumes from the test shed. The noise created by the concurrent running of the unsilenced R and three Kestrels was described by Rodwell Banks, the engineer responsible for the fuel mixture used on the R:

Reverberation from walls and roof is such that at certain speeds one cannot keep still: the whole body seems in a state of high frequency vibration. One shouts at the top of one's voice but cannot even feel the vibration of the vocal chords.[4]

It is said that people living up to fifteen miles from the testing area could hear the engine running. It is rather unlikely that such noisy testing would be tolerated or allowed today. Perhaps even more surprising is the fact that if, after normal working hours, problems were encountered during testing of the engine, a message would be flashed onto the screens of Derby cinemas asking members of the Experimental Department to report for work in order to correct any difficulties as soon as possible.

The R engine was indeed successful in powering the Supermarine seaplanes to secure the Schneider Trophy permanently for Britain, winning the 1929 and 1931 races. Soon after the 1931 race, the R powered Supermarine S.6B reached a new world speed record of 407mph. The R engine went on to power racing cars and speed boats, capturing both the world land speed record and world water speed record. The author, Steve Holter, sums up the design of the Rolls-Royce R engine with these words:

> *Quite simply the R-type engine was far ahead of its time, a marvel of British skill and ability.*[5]

The concentrated experience gained from developing the R engine enabled Rolls-Royce to progress confidently with the PV12 engine that was the Merlin's direct parent. The Schneider Trophy was won despite the unwillingness of the Government of the day to support the challenge, and the story of how it was secured for Britain is a remarkable tale.

Schneider Trophy

The *Coupe d'Aviation Maritime Jacques Schneider*, or more commonly the Schneider Trophy, was initially awarded annually to the winner of a race for seaplanes. The competition, with a first prize of £1,000 (approximately

£100,000 at 2015 prices), was inaugurated in 1913 by Jacques Schneider, a French financier, balloonist and aircraft enthusiast. Between 1913 and 1931, the race was held eleven times – latterly at two-yearly intervals. It was a contest based on pure speed held over a triangular course of 175 miles, later 220 miles. The first race was held in France, and thereafter the winning country hosted the subsequent race. Race supervision was by the Aero Club of the hosting country and the *Fédération Aéronautique Internationale*, the FAI. The FAI is the world governing body for air sports and aeronautics world records, and has its head office in Lausanne, Switzerland. Each competing country could enter a maximum of three competitors and were allowed a maximum of three reserves.

The races were very significant in promoting advances in aircraft design and engine development and performance, ultimately bearing fruit in the Allies' fighter aircraft of World War II. The determination to improve aircraft design and engine performance was not shared by all, however, and as the date for the 1931 Schneider Trophy approached, it became increasingly unlikely that the British Government would provide money for a new aircraft to compete. One must bear in mind that the 1930s were a time of worldwide economic depression following the Wall Street crash of 1929, and accordingly the British Government announced in January 1931 that the expenditure of public money to support the contest was not justified. At that time Britain was in a position to win the Trophy outright with three consecutive wins, but the Supermarine Company of Southampton, builders of the world-beating seaplanes that had already won the trophy in 1927 and 1929, could not fund the development of a new aeroplane by themselves. It looked as though Britain would not be able to compete, even though the contest was to be held on home ground on the Solent.

Despite efforts to encourage the Government to change its mind by highlighting the prestige and sales potential generated by success in the Schneider Trophy races, Supermarine and their chief designer, Reginald Mitchell, gave up hope of competing in the 1931 contest. It was at this point that a 'Fairy Godmother' stepped in in the shape of Lucy, Lady Houston. She donated £100,000 (equivalent to £5 million in 2015) as an unsolicited gift. This amount of investment would allow a new plane to be developed and entered into the 1931 competition.

However, by now the Trophy race was just seven months away, and the possibility of Mitchell and Supermarine producing a new S.7 seaplane was slim. Instead they decided to concentrate on making changes to the S.6 design in order to accomodate the new version of the Rolls-Royce R engine, which was now rated at 2,350hp, an appreciable increase on the earlier 1929 version of 1,900hp. The new engine was not without problems, one of which was how to allow for the removal of the extra heat produced by the more powerful engine. A total of 40,000 British Thermal Units of heat had to be continually removed from the engine, otherwise it would quickly overheat and become unflyable – typically 40,000 BTUs is the amount of heat generated by a 7 foot tall gas patio heater. Mitchell overcame this problem by making modifications to the oil cooling system and by lengthening the floats of the plane to provide more area for cooling.

Because of these modifications, Mitchell often refered to his S.6B as a flying radiator.

Two new S.6B seaplanes were built and were given the serial numbers S.1595 and S.1596. The two 1929 S.6 seaplanes were modified to take the new engines and enlarged floats. They were kept as reserves for the race and were renamed as S.6As.

To fly the aircraft in the races, a new RAF High Speed Flight was formed. It was led by Squadron Leader Augustus Henry Orlebar. With his team of pilots, he arrived at RAF Calshot, the seaplane and flying boat station on Southampton Water, in May 1931, ready to begin trials with the new engines and seaplanes. The first aircraft to reach the team at Calshot was a modified S.6A which Orlebar flew for the first time in June. Early faults with the aircraft, including a tendency for the rudder to oscilate causing flutter, were addressed by Mitchell and cured by fitting balances to the rudder. This modification was immediately added to the other aircraft being built for the contest – the second S.6A and the two S.6Bs.

On 21 July 1931, the first S.6B arrived at RAF Calshot. Augustus Orlebar piloted its maiden flight, which did not actually take place on that day because Orlebar could not get the seaplane off the water. The plane would swing violently through 120 degrees, caused by the increased torque from the new engine. Mitchell determined that the answer to this problem was to use a new, larger propeller. Accordingly a 9 foot diameter propeller was fitted and that did solve the problem, but a further problem occurred with the upgraded engine when engine splutter forced the test pilot to land. Upon inspection Mitchell found that the fuel filter had become blocked by surplus jointing compound coming loose, although the compond actually in the joints was perfectly sound and doing its job without becoming loose. The filter was cleaned, then Mitchell insisted that the planes be flown until all the surplus compound had been removed from the engine through successive cleaning of the fuel filter. John Boothman of the High Speed Flight, who was eventually to become Air Chief Marshall Boothman, later recalled Mitchell's initial answer when the splutter problem first arose: "You will just bloody well have to fly them until all that stuff comes off."[6]

They did, and it worked.

Hopes of an exciting, competitive race faded when France withdrew her entry because her seaplane would not be ready in time, followed by Italy's withdrawal in August after the death of her star pilot, Flight Lieutenant Monti. He died when test flying a new high speed seaplane that crashed in Lake Garda. A postponement request from both France and Italy was refused by the British Royal Aero Club.

It was feared there would be little interest in the race without any foreign challenge, but interest was very keen, and one estimate gives a spectator figure of 500,000. The 1931 Schneider Trophy races began on 13 September. It was a foregone conclusion that Britain, being the only country left in the competition, would win. Sure enough, RAF pilot John Boothman lapped the 30 mile triangular course seven times at an average speed of 340mph in a S.6B, thereby winning the Schneider Trophy outright for Great Britain. Huge crowds watched the races, particularly around Southsea, and the course of the race went very close to West Wittering, home of Sir Henry Royce. It would be nice to think that Sir Henry was in his garden watching the seaplanes power overhead, using the engine that he and his engineers had been instrumental in creating years before with their sketches in the sand on the beach nearby.

Later, on 29 September, another member of the High Speed Flight, George Stainforth, flew an S.6B at an average speed of 407mph. He thereby became the first human to exceed 400 mph. A short profile of Stainforth and Orlebar is included in Chapter 9. These brilliant pilots helped realise the potential of Mitchell's design genius and the undoubted superiority of the R engine.

The last *Coupe d'Aviation Maritime Jacques Schneider* was over, and thanks to R.J. Mitchell, Supermarine, Rolls-Royce, the pilots of the RAF High Speed Flight, and especially Lucy, Lady Houston, the Trophy would remain in Britain forever. Today it sits proudly in the Flight Exhibition Hall at the London Science Museum, along with the Supermarine S.6B.

Captain Ellison Hawkes, veteran of World War I, prolific author of popular science books, and lecturer, wrote in his expert account of the Schneider Trophy races:

In retrospect, it is seen that the Schneider Trophy Contests played a vital part in the history of Great Britain – even, it may rightly be said, in the history of the world – for it was as a result of the experience and incentive they provided that the Spitfire came to be evolved. Had they [and the hurricane fighter] not given us our first victory in that grim episode, without doubt Britain would have been invaded and

> *– as we were unprepared and unarmed – the whole history of civilisation might well have been changed from what it is today.* [7]

The result of the "experience and incentive" was also to provide the impetus for the rapid engine development that resulted in the Merlin. As the Managing Director of Rolls-Royce at the time, Arthur Sidgreaves commented in 1931 to those who wondered what was the value to the motor and aviation industries of winning competitions like the Schneider Trophy races:

As a result of the tests this year all the main components of these engines have undergone a definite improvement, and in consequence the life of the standard engine in service will be much longer than it would otherwise have been.

From the development point of view the Schneider Trophy contest is almost an economy, because it saves so much time in arriving at certain technical improvements. It is not too much to say that research for the Schneider Trophy contest over the past two years is what our aero engine department would otherwise have taken six to ten years to learn.[8]

PV12 Engine

Following the success of the R engine in the Schneider Trophy races, it was clear that Rolls-Royce had the potential to produce an engine that would be very useful to the Royal Air Force. Again Government assistance was not forthcoming at that time. Undeterred, Rolls-Royce went ahead with development of what was called the PV12 engine in the expectation that a very good engine would lead to orders from the Air Ministry and that development was in the national interest. PV stood for Private Venture as Rolls-Royce received no Government money to develop the engine during this initial phase; 12 referred to the fact that it had twelve cylinders.

The basic premise of the PV12 was to produce an engine of similar performance to the R engine, though with a much longer life. The decision to press ahead with development of the PV12 was taken by Sir Henry Royce in October 1932. Sadly Sir Henry died in April 1933 and was never to see the consequences for the world of his decision.

Rolls-Royce had realised that an engine with more power would be needed if its leadership in aero engine manufacture, regained through the Kestrel, was to be maintained. The PV12 was the answer. It was developed from the Schneider Trophy winning R engine and was a major improvement on the Kestrel engine. The PV12 engine was first run on Sunday 15 October 1933, and it passed its type test – that is its formal testing and approval for flight by the relevant authority – in July 1934. During these early test runs it generated 790hp and was first used in a Hawker Hart biplane in early 1935. What was needed now from the Government was backing and funding to further develop the PV12. When that was secured it could follow the convention and be renamed after a bird of prey.[9]

Chapter 2
The Merlin

Background History

In 1935 the Air Ministry issued a directive requiring a new fighter aircraft that could attain a speed of 310mph. Supermarine and the Hawker Company responded to this new requirement and developed their respective fighter aircraft, the Spitfire and Hurricane, around the Rolls-Royce PV12. In 1936 both Supermarine and Hawker received firm orders from the Air Ministry to manufacture their fighter aircraft in large quantities. This led to the PV12 receiving Government funding and corresponding orders. The PV12 engine was then renamed the Merlin.

The Merlin was a liquid-cooled V-12 in-line aero engine, having two banks of cylinders arranged in a V configuration, with less than 180° between each bank, driving a common crankshaft. An in-line aero engine has its cylinders arranged in a line similar to many automotive engines. Its main advantages are its narrowness and corresponding small frontal area allowing the aircraft, typically a single engine fighter, to have a slender front fuselage. This improves the aerodynamics, with the additional advantage of improving the pilot's vision. The airflow around an in-line engine is not as good as a radial engine, however, and because of this, liquid cooling of the power unit is required. All Merlin engines were 'right hand tractor' in that the propeller rotated clockwise viewed from the pilot's seat.

The Merlin had a displacement of 27 litres. Engine displacement, or capacity, refers to the volume swept by all the pistons in the engine, and does not include the volume of air above the piston where the initial spark fires. Each piston compresses air within a cylinder, while fuel, typically petrol, is injected into the compressed air and ignited by a spark plug. This ignition further increases the air pressure within the cylinder. This high temperature, high pressure gas pushes the piston down. The distance the piston moves is called the stroke. The piston is directly connected to the transmission which turns the propeller of the aircraft.

The early Merlins were not without quality problems that led to questions about their reliability. They had a tendency to leak coolant and they often showed excessive wear to the camshafts and crankshaft main bearings. The cylinder head also frequently cracked. When the engine reached modification version F, however, all major quality concerns had been resolved, and version F became the Merlin Mark I.

During the Battle of Britain the majority of Spitfires were fitted with the Merlin Mark III of 1,030hp, because at that time the Mark III was readily available. As the engine further developed, the Mark XII Merlin was able to generate 1,150hp and was used in the Spitfire Mark II. The Mark II Hurricane fighter used the Merlin Mark XX which generated 1,480 hp. The Merlin Mark 45 generated 1,515hp and was used on the Spitfire Mark V; by the end of the war this had become the variant of Spitfire produced in more numbers than any other model. One can see from these ever increasing figures for horsepower output how the Merlin was disposed to power increases. The potential of the engine was supremely demonstrated very early on, before the onset of war, when in 1937 a highly modified experimental Merlin generated over 2,100hp on test.

The Merlin Mark XX was the engine used on the first operational versions of the Avro Lancaster bomber. In combination, the four Merlins on those early versions of the Lancaster gave it a total power output of 5,920hp. There were 29,508 Merlin engines produced for the Lancaster, more than for any other aircraft used by the RAF.

In some regards, it is surprising that the Merlin was designed at all. It was originally perceived as something of a stop-gap engine to fill a void in production between the Kestrel and the Vulture. A relatively simple development of the Kestrel called the Peregrine had promised to be very successful, and a twenty-four cylinder variant named the Vulture was expected to give 1,700hp. The Vulture had a troubled time during its development, however, and two aircraft programmes that were to use it as their power plant, the Avro Manchester bomber and the Hawker Tornado fighter, were both cancelled. Other engine manufacturers, most notably D. Napier and Sons, had major engine projects that included the Dagger and the Sabre. Both had problems, and only the Sabre became available for use in 1942 to power the Hawker Typhoon.

The other major aero engine manufacturer, Bristol, did

not produce in-line engines, preferring instead to design and build radial engines, which were the next most common aero engine configuration used during World War II. In a radial engine the cylinders point outward from a central crankshaft in a similar way to the spokes of a wheel, resembling a star when viewed from the front – and indeed this type of power plant is called a star engine in some countries. A successful example of the radial engine is the Bristol Hercules engine of which over 57,000 were made during and after World War II. Hercules engines powered the Bristol Beaufighter and many other aircraft. It was the Merlin, however, that provided all of Britain's in-line aero engine requirements during most of World War II.

What is sometimes forgotten about the Merlin when compared with the other competing engines used during World War II is its compactness. The Merlin had a capacity of 27 litres, whereas the German engines, like the BMW 801 engine in the FW190, had 42 litres; the DB601 in the Me109 had 39 litres; and the Jumo 211 in Heinkel bombers had 35 litres. It might seem likely that the bigger the capacity of the engine, the greater its power; sometimes this is so, but other factors come into play that affect power output: for example, the use of superchargers to force more air into the cylinders. This was the case with the Merlin, which always punched above its weight because of the superior engineering employed in its development.

The Supercharger

Rolls-Royce engineer Cyril Lovesey began working with others on developing the Merlin in the 1930s, and in 1940 was placed in charge of the engine development programme. In 1946 Lovesey delivered a lecture on Merlin development. In this lecture he explained in detail the major factors that made the Merlin such a great engine. He divided the improvements to the Merlin into three classes:

- improvement of the supercharger,
- improved fuels,
- development of mechanical features to take care of the improvements afforded by 1 and 2.

Lovesey explained that the power output of an engine largely depended upon the mass of air it can be made to consume efficiently, and in this respect the supercharger played the most important role. An engine subjected to increased power output from using a supercharger has to be capable of dealing effectively with the following:

- greater mass flows with respect to cooling the engine,
- freedom from too early detonation of the fuel/air mixture which would cause misfiring of the engine,
- ability of the engine's component parts to withstand higher gas and inertia loads.

This is why point 3 above – development of mechanical features to take care of the improvements afforded by 1 and 2 – was relevant with the requirement of greater cooling ability to deal with the increased heat generated by the engine, reducing the probability of the more volatile fuel detonating too early, causing the engine to misfire, and the possibility of failure of individual engine components due to the increased loads and speeds.

As the Merlin developed, so did the supercharger, which fitted into three broad categories:

1. Single-stage single-speed gearbox: Merlin I to III, XII, 30, 40, and 50 series (1937–1942).
2. Single-stage two-speed gearbox: experimental Merlin X (1938), production Merlin XX (1940–1945).
3. Two-stage two-speed gearbox with intercooler: mainly Merlin 60, 70 and 80 series (1942–1946).

The first Merlin supercharger allowed the engine to reach maximum power at 16,000ft. It was located at the rear of the engine, and because of this was of an inefficient and unusual design that limited the supercharger's performance. A new design by Dr S. Hooker increased the maximum power obtainable from the engine to an altitude of 19,000ft. This eventually led to the development of the single stage Merlin XX and 45 series. Further developments by Dr Hooker saw the introduction of a two-stage supercharger on the Merlin 60 engine. Thus fitted, the two-stage two-speed supercharged engine gained an extra 300hp at 30,000ft.

The Merlin 45 was used in the Spitfire Mark V from October 1940. This was produced in the greatest numbers of any Spitfire variant. The Merlin 61, used in the Spitfire Mark IX from July 1942, was the second most produced variant. This Mark of engine also became the power plant of the North American Mustang fighter, which will be discussed in detail in a later chapter.

Another of the superior engineering developments incorporated into the Merlin was the use of a pressurised engine cooling system.

The Cooling System

The Merlin's liquid cooling system was to be provided initially by condensers in the aircraft's wings, together with a small retractable radiator for use at low flying speeds. Eventually, in order to gain more power from the engine an alternative cooling agent was employed. This new agent was Ethylene Glycol, which is more efficient than water at removing heat. Its use allowed the removal of the condensers in the wings. Ethylene Glycol had the disadvantage of being flammable, however, and this increased the risk of an aircraft fire. To overcome this disadvantage, a mixture of Ethylene Glycol and water under pressure was introduced. The fire risk was reduced, and now the Merlin had a pressurised cooling system.

A third important factor that allowed the Merlin to develop more power and maintain its 'one step ahead' status was the use of 100 octane fuel.

One Hundred Octane Fuel

The octane rating or number is a standard measure of the performance of an aviation fuel. The higher the octane number, the more compression the fuel can withstand before igniting. Fuels with high octane ratings are usually used in high performance petrol engines that require larger compression ratios. By contrast, fuels with lower octane numbers are ideal for diesel engines which do not compress the fuel. They compress the air, and then inject the fuel into the air heated by the compression. Petrol and aviation fuel engines rely on ignition of air and fuel compressed together until the end of the compression stroke, where the ignition is carried out by using a spark plug. High compressibility of the fuel is very important in high performance petrol engines such as the Merlin.

The Merlin was able to take advantage of the new 100 octane fuel developed in the United States. The Air Ministry was convinced that 100 octane fuel could boost the power of aero-engines, and began importing small quantities from the United States, sending them to Rolls-Royce for testing in the Merlin. The first consignment of BAM100, British Air Ministry 100 octane fuel, was shipped in June 1939. The Ministry began stockpiling the fuel, while the RAF continued using standard 87 octane fuel. Most shipments came from Esso and Shell refineries in the United States, but when war was declared in September 1939, the US Congress invoked the Neutrality Act which prohibited the export of strategic materials to belligerent nations. Fortunately some shipments came from Curacao and Aruba, islands in the Caribbean Sea where there were major oil refineries, so valuable shipments continued supplying 100 octane fuel to Britain and the RAF.

In the first half of 1940, all Hurricane and Spitfire fighters were transferred to 100 octane fuel. Bill Grunston, author of many books on aviation and the military, said of the difference 100 octane fuel made to the power of the engine:

Instead of being limited to 6lbf/sq.in boost, pilots could go to full throttle and 12lbf/sq.in boost, thus increasing the power of the Merlin from 1,000hp to 1,310hp, allowing the aircraft to fly higher and faster.[10]

The importance of 100 octane fuel was further underlined by the fact that the German Messerschmitt Bf109E fighter was lighter than either the Spitfire or Hurricane and had an engine of greater capacity (as stated previously, 27 litres in the Merlin compared with 39 litres of the DB601 engine in the Me Bf109E). As there was little difference in the weight of the competing engines, it was vitally important that Rolls-Royce kept the Merlin ahead of the German engines through increased tolerance to higher manifold pressures for more power at low altitudes, and also by superior supercharging for greater power at high altitudes. The Merlin succeeded in both respects.

All this continuous improvement and development was not without some failures. Each one of them required extensive investigation to overcome the engineering problems and move forward.

Failure Investigation

The Rolls-Royce Failure Investigation teams had a challenging philosophy regarding engine failures and setbacks of whatever kind. All new engines have teething problems. Once these are ironed out there is an ongoing need to ensure any further engineering concerns caused by the constant development and power increases of the engine are addressed and cured. It is how these problems and concerns are overcome that contribute greatly to the ongoing success of the engine. Rolls-Royce had a particular philosophy in relation to the many engineering setbacks that arose during the development of the Merlin. The Failure Investigation Department had one overriding rule:

There is no such thing as an isolated failure. The isolated failure of today is the epidemic of tomorrow.[11]

This meant that every new failure had to be investigated and acted upon, no matter what the previous history might be. In this way, the Failure Investigation team ensured that the Merlin never failed for the same problem more than once.

The engineer in charge of the Merlin Engine Failure Investigation team during World War II was Alec Harvey-Bailey. His role involved the progress of engineering improvements and the repair of damaged or worn engines. Harvey-Bailey would often, where necessary, visit active RAF squadrons and fly aircraft in order to become familiar with particular problems. This gave him a greater insight into the causes of failures and how best to correct them.

Development History
The design and development work carried out on the Merlin in order to provide the Allies' aircraft with the necessary superiority produced twenty-one different power ratings and fifty-two different marks of engine, all in order to meet the installation needs of a diversity of aircraft ranging

from the single-engine Spitfire, Hurricane and Mustang, through the two-engine Mosquito, to the four-engine Lancaster. In addition to this development programme, the Merlin was also subject to over 1,000 engine component modifications aimed at improving reliability, enabling easier servicing and simplifying production. To list all modifications and marks would turn this account into a mighty tome, and accordingly this section is confined to a brief overview of the development of the engine. The main modifications to the Merlin that produced these increases in performance have been discussed earlier in this chapter. The information in the Glossary will also be helpful to the reader.

Between 1939 and 1945 the Merlin saw operational service in nineteen different types of aircraft. The complete range of aircraft, including development aircraft and other types that did not enter full scale production, is listed in Addendum A.

Engine ratings in operational service progressed as follows:

- 1940 – 1000hp at 16,000ft – Merlin III,
- 1942 – 1000hp at 30,000ft – Merlin 61,
- 1944 – 1000hp at 36,000ft – Merlin 113/114.

The above figures are for the Spitfire fighter. Merlin engines in other aircraft types would show similar increases as the war progressed.

The following graphs show the effect of Merlin development on the performance of the Spitfire and Mustang fighter aircraft.[12]

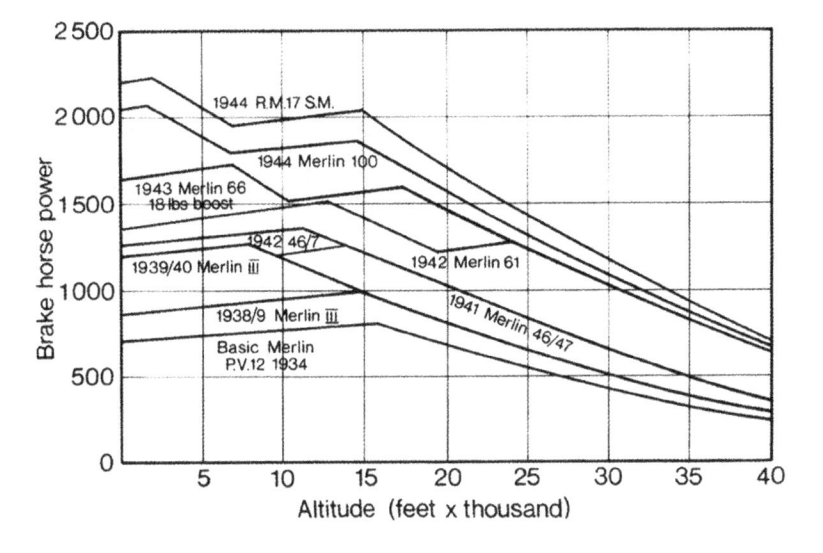

Performance comparison

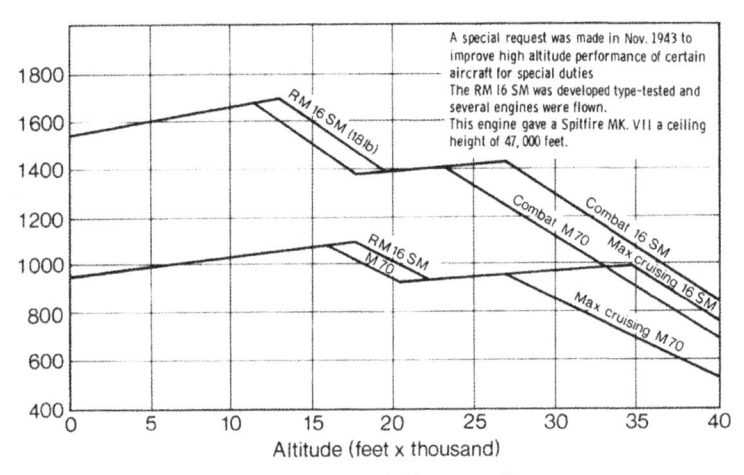

Merlin development to meet special high altitude conditions

A special request was made in Nov. 1943 to improve high altitude performance of certain aircraft for special duties. The RM 16 SM was developed type-tested and several engines were flown. This engine gave a Spitfire MK. VII a ceiling height of 47,000 feet.

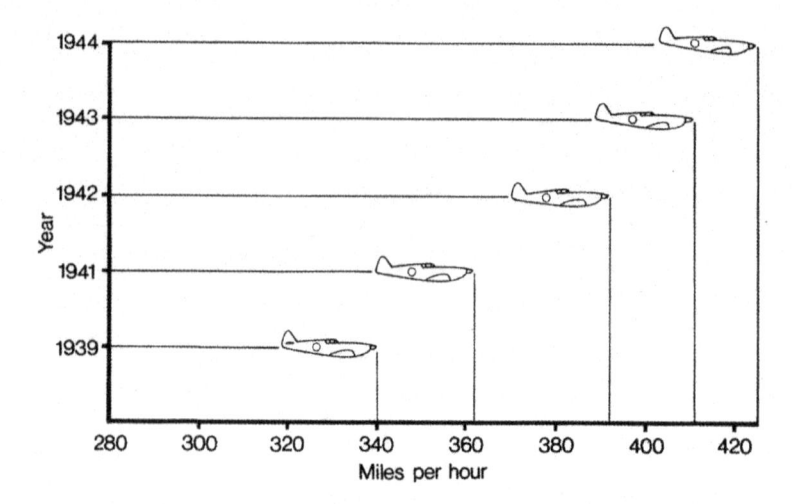

Spitfire speed increase due to Merlin development

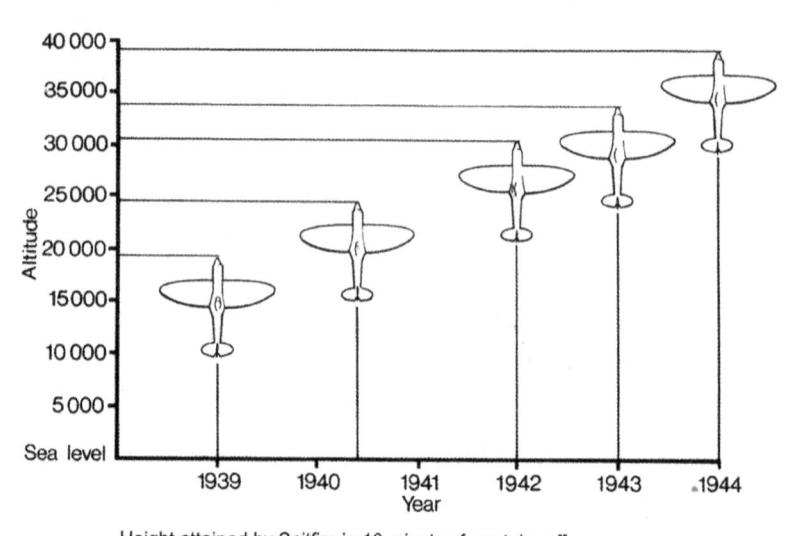

Height attained by Spitfire in 10 minutes from take-off

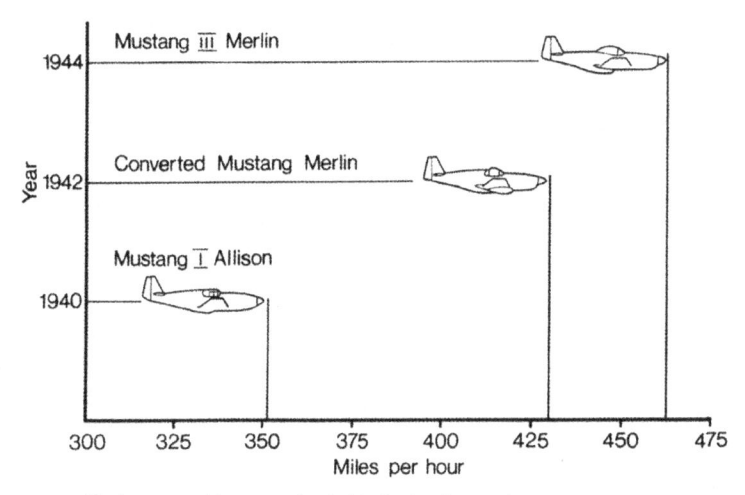

Mustang speed increase due to Merlin development

The Difference

What exactly was so special about the Merlin that it should hold such an exalted position in aviation history and the history of World War II? From the books, articles, stories and comments concerning the Merlin consulted during research for this book, one fact stands out above all. The Merlin was an engine that had the development potential to be improved and upgraded beyond any other engine existing at that time, whether of the Allies or the Axis side. Throughout its life, every part of the engine was changed, modified, improved and enhanced to the nth degree. Two books that document some of the modifications and detail the marks of engine are The Merlin in Perspective – the combat years by A. Harvey-Bailey and Rolls-Royce Piston Aero Engines by A. Rubbra (see Bibliography). Both are well worth reading for those interested in the extraordinary lengths to which great engineers will go in order to gain advantage.

In summary, all those engine improvements meant that by the end of World War II the Merlin engine was producing more than twice the power of its original, resulting in the Spitfire's maximum take-off weight being doubled, its rate of climb more than doubled, and its maximum speed being increased by a third. All this in the same airframe. An incredible feat of development, and a tribute to the brilliance of R.J. Mitchell's original Spitfire design.

Following the Battle of Britain, Spitfires fought in every operational theatre of World War II and remained in RAF service until 1954.

To quote Squadron Leader Nigel Rose, Spitfire pilot in the City of Glasgow 602 Squadron, during his interview with the *Herald of Scotland* newspaper in 2010 at the time of the seventieth anniversary of the Battle of Britain:

I flew Kestrels in Hawker Harts and Hinds early on and you really noticed the difference in power [when using the Merlin]. *I flew 900 hours in Spitfires during the war and never had any trouble at all. All through the war the German planes, the Messerschmitt's and so on, and British planes were stepping up each other in performance, manoeuvrability and speed. The Merlin kept up, it was improved, and it got more powerful.*[13]

A connection has been drawn between the various Rolls-Royce engines under development and manufacture from 1915 to 1945 in order to show a direct line of improvement and succession from one engine to another through the passage of time. This has proved difficult. While it may be said that the Eagle and the R engine contributed much to the design of the Merlin, and the PV12 may be the Merlin's direct parent, engine development is much more complex. All the in-line aero internal combustion engines mentioned in this text have their connections. All, to varying degrees, have contributed to the achievements of their successors. Indeed, many of them shared very similar design teams. With the Merlin, the Allies were very fortunate because here was an engine that was capable of almost endless improvement, stayed ahead of the opposition throughout the war and was the best in the world at the time. Although the Griffon engine overtook the Merlin in performance towards the end of hostilities, it was not until the advent of the Gas Turbine engine that the Merlin was appreciably overtaken in performance.

Note: A cutaway diagram of the Merlin engine is included in Addendum B.

A Perspective

Mike Evans, author and the founder of the Rolls-Royce Heritage Trust, put things very succinctly when he said:

> *Without the Merlin, we would not have won the Battle of Britain and Hitler may have crossed the channel.*[14]

In 1945, the years of constant development on the Merlin stopped. Had the war not ended then, more horsepower and more improvements would have been available. Rolls-Royce had schemes ready to improve further the air entry to the supercharger using a three-speed supercharger that had been fully tested and was ready to be incorporated. Merlin milestones include the fact that it was the only British engine to be used by the USA Air Force in the Mustang long-range fighter. (The only British engine used in World War II, that is. The Pegasus engine for the Harrier jump jet and its American equivalent, the AV8-B, and the English Electric Canberra bomber, powered by the Avon gas turbine, are two other British engines used by the air forces of the USA). The Merlin powered more than 50,000 single engine fighters, and after the war was used to power the Lincoln bomber, a derivative of the Lancaster, which was the mainstay of RAF Bomber Command until 1955.

The power output of the Merlin had reached 2,000hp plus by the end of hostilities. What began with a 50hp car engine in the 1908 Silver Ghost had gone through the Eagle at 300hp and the PV12 at 1,100hp, peaking in the racing R engine for the Schneider Trophy at 2,800hp. These engines all contributed something to the Merlin, though, as already stated above, they are not alone. Many other developments and modifications, large and small, came as a result of knowledge gained from work on other engines, including the Buzzard, Goshawk and Vulture.

In summary, between 1939 and 1945, the Merlin was used in a great many different types of operational service aircraft (full listing of aircraft types that used the Merlin are in Addendum A). There were Merlin variants with single-stage single-speed superchargers, single-stage two-speed superchargers, and two-stage two-speed superchargers with intercooling. Power ratings ranged from 1,000hp to over 2,000hp.

It should also be remembered that the Germans did eventually, by mid-1944, have a fighter that could outstrip the Spitfire, Hurricane and Mustang in every department. This aircraft was the Messerschmitt ME262 which was the world's first operational jet-powered fighter aircraft. The ME262 had a top speed of 560mph and a service ceiling of 37,500ft. It was used in a variety of roles, including as a light bomber, reconnaissance and fighter aircraft. By the time of its introduction, the Allies had an overwhelming command of the air, having much larger numbers of fighters and bombers than the Axis powers. Even though the ME262 was faster than any Allied plane, it had little effect on the outcome of the war because it came into full service far too late and in comparatively small numbers. Of course, had the ME262 been available much earlier than it was this history would have been considerably different from the one you are reading now.

Rolls-Royce's great contribution to the ability of the Allies to overcome the enemy air force and strike at the heart of Germany was the Merlin engine, the Griffon engine providing a back-up during the later stages of the conflict. While the other engines mentioned gave much useful information and enabled different design philosophies to be evaluated, it was ultimately the Merlin that was the stronghold of the Allies' wartime piston-engine success.

The Merlin was not without competition, of course, as Rolls-Royce was not the only British manufacturer of aero engines. Indeed, the Napier Company and the Bristol

Engine Company provided the RAF with invaluable support through the supply of engines that powered a number of frontline aircraft. This was the friendly competition. Daimler-Benz, BMW and Junkers Motoren Engine Company provided the unfriendly competition.

Chapter 3
The Competition

Friendly
Griffon Engine

The Rolls-Royce successor to the Merlin was the Griffon which entered service with the RAF late in the war. It was, in many ways, of an older design concept, being based on the Buzzard engine which first ran in 1928. Work on the Griffon began again in earnest in 1939, but in early 1940, on the orders of Lord Beaverbrook, then Minister of Aircraft Production, it was halted temporarily in order to concentrate all efforts on the output of Merlins.

The first of the Griffon-engine Spitfires flew on 27 November 1941. Although these were never produced in the large numbers of the Merlin-engine variants, they were an important part of the Spitfire family, and in their later versions kept the Spitfire at the forefront of piston-engine fighter development. The first Griffons had single-stage superchargers and were fitted to the Spitfire Mark XII. These aircraft arrived just in time to take on the Focke-Wulf 190 fighter bombers that were attacking England's south coast. For high altitude flying, a two-stage supercharger was introduced in the Griffon, and these were fitted in the Spitfire XIV and XVIII. This enabled the Spitfire to stay in the forefront of fighter performance until the end of the war. The Griffon was the last of the V-12 aero engines to be produced by Rolls-Royce. Production ceased in 1955, and by then over 8,000 engines had been manufactured.

Hercules Engine

The Hercules engine was manufactured by the Bristol Engine Company. The engine was of a radial, or star, configuration with fourteen cylinders and sleeve-valves. It is best remembered as the power plant for the two engine Bristol Beaufighter, a heavy fighter used in tactical and anti-shipping roles throughout the war, when 5,928 were built.

With the advent of full scale Lancaster bomber production, the extra demand placed on Merlin supplies raised the possibility of a bottleneck occurring. Accordingly, it was decided to fit the Mark II Lancaster with the Hercules radial engine. The first flight test of a Hercules powered Lancaster took place on 26 November 1941, and the results were good. The Lancaster production lines at Avro were full to bursting, so a contract with Armstrong-Whitworth was negotiated, and between September 1942 and March 1944 they built 300 Mark II Lancasters at their Baginton factory near Coventry, all powered by the Bristol Hercules Mark XVI engine. The Lancaster Mark II had its first official operational raid on 16 January 1943.

When fitted to a Lancaster, the Hercules engine gave a similar performance to a Merlin powered Lancaster in that the maximum speed achieved was 265mph at 14,000ft, with a cruising speed of 167mph. The Mark I Lancaster had a maximum speed of 282mph at 13,000ft and a cruising speed of 200mph. The Hercules engine gave a superior rate of climb when compared with the Merlin, up to 18,000ft, but progress fell off above this height. Merlin powered Mark Is and Mark IIIs were operating at up to 22,000ft. Needless to say, the higher a bomber can fly during operations, the less chance there is of it being shot down by enemy fire. On the positive side, the air-cooled Hercules could withstand a greater degree of damage from the enemy than the liquid-cooled Merlin.

Production of Mark IIs was discontinued in November 1943, when Armstrong-Whitworth switched to Merlin-powered Mark I and III Lancasters. This move was in part due to the demand for Hercules XVI engines to power the four-engine Hadley Page Halifax bomber, of which over 6,000 had been built by the end of hostilities. The Hercules engine was produced in large numbers at Bristol, and over 57,000 had been built before production ceased.

Sabre Engine

The Sabre engine powered the Hawker Typhoon fighter-bomber and Tempest fighter, and was manufactured by the Napier Company in Acton, London. The engine was a twenty-four cylinder H configuration, which meant that the cylinders were aligned so that, when viewed from the front, they appeared to form the letter H. This formation allowed the building of multi-cylinder engines that were shorter than the alternatives.

Problems arose with the Sabre engine as soon as volume production started. The Sabre engines used for early testing had been hand built and carefully assembled by top craftsmen.

Production engines did not have this luxury, relying on the interchangeability of engine components required for mass-production. This resulted in many mechanical problems as the engines entered squadron service. The engine also lacked performance at altitude, and there were a number of forced landings recorded due to engine problems.

The Sabre engine became more reliable post-war, and had evolved to become one of the most powerful in the world by the end of hostilities, when it generated 3,500hp in late model prototypes. Napier test ran a Sabre at a colossal 4,000hp towards the end of the war. No other engine equalled this performance. By the end of its production life, over 8,000 Sabre engines had been produced.

Unfriendly

The question is often asked as to how the Merlin compared with enemy front line equipment. A book could be written on the subject.[15]

Well, this is not that book. However a brief overview of the relative performance of the Merlin and the Axis engines would, I'm sure, be worthwhile to discuss. The three main Axis engines of World War II were manufactured by Daimler Benz, BMW and Junkers. All these engines were appreciably larger than the Merlin, which had 27 litres capacity compared with the BMW at 42 litres, the Daimler-Benz at 36 litres and the Junkers Jumo 211 at 35 litres. All three engines, however, were out-performed by the Merlin over the course of the war.

As an example of this better performance, the Merlin Mark 61 in the Spitfire IX, introduced in March 1942, prompted Battle of Britain Ace Johnnie Johnson to say:

The Spitfire IX was the best Spitfire. When we got the IX, we had the upper hand, which did for the 190s.[16]

The Mark 61 engine had the supercharger technology which was being constantly developed by Rolls-Royce engineer Stanley Hooker and his team, including Geoffrey Wilde and Cyril Lovesey. This technology kept the Merlin ahead of the opposition, and as a result the Spitfire IX was able to achieve and maintain superiority over both the FW190A and the new Bf109G.

Junkers Jumo 211 Engine
The Jumo 211 engine was an inverted V-12 in-line petrol injection power plant best known for being used primarily in bombers, including the Heinkel He111 and the Junkers Ju87 and Ju88.

Inverted inline aircraft engines are installed in airframes upside-down, such that the crankshaft is at the top of the engine and the cylinder heads are at the bottom. The main advantages of this are improved visibility for the pilot flying in a single seat fighter, and improved access to cylinder heads and manifolds for the ground crew, allowing easier servicing of the engine.

The Jumo 211 was the most produced Axis engine of World War II, with over 68,000 manufactured.

Daimler Benz DB605A-1 Engine
The Me109 was powered by the Daimler-Benz DB605A-1, a liquid-cooled inverted V12 in-line engine. During the early periods of fighter combat, the Spitfire had the performance

edge on the Me109 at 20,000ft, again mainly due to Stanley Hooker's supercharger technology and the use of improved fuel. However, when Germany abandoned the tactic of escorted bomber formations and resorted to high-flying fighter bombers, it became evident that at around 30,000ft the Me109 was quicker and had a faster rate of climb than the Spitfire. The Daimler Benz DB601 engine, fitted to the Me109, had more power than the Merlin Mark III at that altitude, giving the German fighters a definite advantage. During 1941, the Merlin caught up following further improvements to the supercharger and other components which, when fitted in the Spitfire V, brought it back on equal terms with the Me109 regarding speed and rate of climb.

BMW 801 Engine

The BMW 801 is best known as the engine that powered the Focke-Wulf FW190. It was an air-cooled fourteen cylinder petrol injected radial engine used in a number of Luftwaffe aircraft of World War II. Production versions of the twin-row engine generated between 1,540-1,970hp. Its performance against the Merlin was a succession of one engine out-performing the other and then the reverse. Initially the Merlin powered Spitfire had the edge on the Fw109, then at the end of 1941 the FW190 got ahead of the Spitfire in speed and rate of climb, where it had gained an advantage of some 250hp over the Merlin 45/46 in the Spitfire at 16,000ft. The Merlin soon caught up and overtook the BMW 801, and by June 1942 the Merlin 61 in the Spitfire IX was outperforming the FW190 at altitudes of around 25,000ft.

Chapter 4
Work Until it Hurts

Although many Merlin engines were made in Derby before the onset of hostilities in 1939, the average figure for production of the Merlin during World War II was approximately eighty engines every day. This figure takes into account all the different marks of engines manufactured over the almost six-year course of the war, providing a total of 168,040 engines. At the beginning of the war, one factory in Derby was producing Merlin engines. By the end of the war in 1945 there were five factories producing Merlins, and the probable production rate then would have been in the region of two to three hundred engines per day. Indeed the Rolls-Royce Hillington factory near Glasgow produced, at its peak of output, a hundred engines in one day.

How was this truly prodigious feat of organisation and manufacturing production achieved? How was this trickle of engines turned into a torrent of power?

The Factories That Produced the Merlin Engine
Of the five separate factories producing the Merlin engine at the end of the war in 1945, four were part of the wartime

'Shadow Scheme'. In 1936, during the build-up to World War II the British Government realised that the Royal Air Force must expand quickly to meet the possible challenges of a world war. The Secretary of State for Air, Sir Kingsley Wood, made immediate plans to extend the existing factories producing aircraft and aero engines and to set up new factories to help achieve the required output. This became known as the 'Shadow Scheme'. The project was headed by Herbert Austin, and the new factories were known as 'Shadow Factories' because they were following the practices of existing factories and learning how to produce their products. Government grants were made available to facilitate the scheme, and it soon became obvious that factories producing motor vehicles were the easiest to convert to the mass-production methods required for aircraft and aero-engine manufacture. Rolls-Royce was directly involved in this scheme, and a new mammoth manufacturing organisation was developed to produce the Merlin engine in very large numbers.

The following gives an overview of the factories involved in producing the Merlin. (The Derby factory was not part of the Shadow Factory system.)

Rolls-Royce Derby

The existing Rolls-Royce facilities in Nightingale Road, Derby, were not suitable for the large scale engine production necessary for the war effort, although the factory had increased in size by twenty-five per cent between 1935 and 1939. The Derby factory had been built and developed primarily for producing high-class engineering products in small quantities. Its structure required a large proportion of highly-skilled men necessary to produce the required standard of engineering, coupled with the ability to change or modify the product with the minimum of fuss or delay. In short, the Derby factory was a very large development factory and not ideally suited to the large scale manufacturing required during wartime. Nevertheless, it was planned to build the first two or three hundred Merlin engines there and resolve any engineering teething troubles before commencing further large scale production at other shadow factories planned in order to ramp-up production of the Merlin. The Derby factory had a workforce consisting mainly of design engineers and highly-skilled production workers, the large majority of whom were men, and carried out most of the development work on the Merlin, with flight testing on the engine held at nearby RAF Hucknall.

The Derby factory produced 32,377 Merlin engines before production ceased after the end of the war. All the Merlin engines used during the Battle of Britain were manufactured in Derby. After the war the factory was responsible for developing and manufacturing gas turbine engines, including the RB211 engine which was used primarily in the Boeing 747, and the Tay series of engines for the civil aerospace market that included powering the passenger aircraft manufactured by the Airbus Company. The Derby factory closed in March 2008 when manufacture of aero engines was transferred to other Rolls-Royce factories around the world.

Rolls-Royce Glasgow Hillington

As it became increasingly apparent that war with Germany was very likely, and with the Government demanding a larger output of armaments from British factories, E.W. Hives, the General Manager of Rolls-Royce, recommended that a factory be built near Glasgow to take advantage of the ample local work force and the good supply of raw steel and metal forgings from Scottish manufacturers. This Government funded and operated factory was built at Hillington. Construction started in June 1939. Workers began moving into the new premises in October 1939, the factory becoming fully operational by September 1940.

The Hillington factory was one of the largest industrial operations in Scotland with 16,000 employees. Unusually, it produced almost all the components required to assemble the Merlin in-house. Other Rolls-Royce factories, including the Derby and Crewe plants, relied significantly on external subcontractors to provide substantial quantities of components for the Merlin. Completed engines began to leave Hillington in November 1940, and by June 1941 monthly output had reached two hundred, increasing to more than four hundred per month by March 1942. A total of 23,675 Merlin engines was produced by the Hillington plant. The record for production is reported to have been 100 engines in one day.

Not surprisingly, a working week of at least seven twelve-hour shifts, the mental and physical demands of wartime on the workforce, as well as the frequent bombings from the German Luftwaffe, resulted in an increase in worker sickness and absenteeism. It was eventually agreed with Rolls-Royce management to cut the punishing working hours to eighty-two hours per week, with one half Sunday per month given as a holiday.

During the first six months of operation, the Hillington

plant took on and trained thousands of machine operators who were to make the many and varied component parts for the Merlin. A number of problems had to be overcome to achieve this dramatic increase in output. First of all, most of the fit men had been called up to fight the war. Only the men who were unfit, too old, or conscientious objectors and, of course, women were available for employment. Most Clydeside factories had agreements with management that only time-served skilled men would operate machine tools. Because of the limitations caused by these agreements, it was necessary to relax the existing practices, and a 'Relaxation Agreement' was drawn up with the Trades Unions to operate for the duration of the war. This would allow the use of unskilled labour, mostly women, for both operating and setting the complex machine tools necessary for production. Soon ninety per cent of machine operators were female, and in spite of the fact that Clydeside apprenticeships were, pre-war, of five years duration, the female operators were trained using the following system:

- first week – new operator observed trained operator by standing alongside while they worked,
- second week – new operator allowed to operate the machine herself,
- third week – new operator became an instructor and taught another girl how to operate machine.

It is notable that the scrap rate (the number of defective unusable component parts produced) for the first six months of operating this system was rather excessive, but this scrap rate improved considerably to manageable levels as the workforce became more used to producing component parts. After the war, the factory repaired and overhauled both Merlin and Griffon engines in addition to producing spare parts.

Requirements for the Avon Turbojet for the Korean War meant that production switched totally to gas turbine engine production by the early 1950s. In 1965 Hillington became a specialist compressor component manufacturing facility for Rolls-Royce. The Hillington factory was closed in 2005, with work being transferred to the more modern Rolls-Royce facilities in Scotland at Inchinnan and East Kilbride.

Rolls-Royce Crewe

Rolls-Royce started building work on a new factory at Crewe in May 1938. The Crewe factory had convenient road and rail links to their existing facilities at Derby. Production at Crewe was originally planned to use unskilled labour in-house and large numbers of sub-contractors to supply complex components, with which Rolls-Royce felt there would be no particular difficulty. However, the number of required sub-contracted engine parts, such as crankshafts, camshafts and cylinder liners, eventually failed to meet both quality and output requirements, and the factory was expanded to manufacture these parts in-house. The first Merlin engines left the Crewe factory production line in 1939.

Use of unskilled labour caused some problems, and withdrawal of labour through strike action by the skilled workforce was not unknown. In 1940 a strike took place at the Crewe factory as a result of women being assigned to replace men on capstan lathes. The workers' union insisted this was a skilled job and they claimed that skilled labour had been displaced. Management stood firm and the strikers returned to work on 19 April after ten days of action. A peaceful solution was agreed, and the women continued working on capstan lathes – but for the duration of the war only.

The Crewe factory was badly damaged by enemy bombing in December 1940, when a lone Ju88 released two bombs on the factory, killing seventeen workers. As a result of this tragedy, the workforce was issued with tin helmets to help protect them from future bombing raids. However, the bombing had little effect on output. Seven hundred Merlins were despatched in January and February 1941, which was eighty-eight more than in November and December 1940. This rate of production was maintained for the duration of the war, and by the end of World War II 26,065 Merlins had been delivered to the RAF.

Post-war the factory was used for the production of Bentley motor cars, and in 1998 the Volkswagen Company bought both the Bentley marque and the factory. Today the factory is known as Bentley Crewe.

Ford Trafford Park, Manchester

Trafford Park proved a boon for producing the Merlin engine. It was redeveloped from what had been, since 1931, a derelict Ford motor assembly plant. In 1938 the new factory was designed in two separate sections to minimise the probable impact of bomb damage on production. Indeed, the redeveloped Ford Trafford Park factory was bombed only a few days after its opening in May 1941, but production was relatively unaffected. The factory was located close to major transport links, giving easy access for the finished product to be supplied to Metropolitan-Vickers, also located in Trafford Park, for use in the Avro Manchester bomber, and the Avro factory at Chadderton for use in the Avro Lancaster.

The Production Engineers at Ford redrew the blueprint drawings for the Merlin, using their knowledge of mass-producing many hundreds of thousands of automotive components to tight tolerances and high quality. This made the Merlin more suitable for mass production. By 1944 they were producing over 400 Merlins a week. Sir Stanley Hooker considered the output from Trafford Park as "Damn good engines"[17], this comment being mainly based on the fact that the number of rejected engines from Trafford Park was zero.

Ford's investment in machinery and their engineers' mass production know-how resulted in constant improvements in production times and manufacturing processes, dropping the 10,000 man-hours needed to produce a Merlin to 2,727 man-hours. Engine unit cost fell from £6,540 in June 1941 to £1,180 by 1945. This was an astonishing achievement and a tribute to the engineers at the Ford plant. There were 17,316 people working at the Trafford Park plant, including 7,260 women. Merlin production started to run down in August 1945 and finally ceased on 23 March 1946. The total number of Merlin engines produced at Trafford Park was 30,428 between June 1941 and March 1946, only 2,000 fewer than that of the main Rolls-Royce plant in Derby which had been producing Merlins for a much longer period.

In his autobiography *Not Much of an Engineer*, Sir Stanley Hooker said:

> ...*Once the great Ford factory at Manchester started production, Merlins came out like shelling peas. The percentage of engines rejected by the Air Ministry was zero. Not one engine of the 30,400 produced was rejected...yet never have I seen mention of this massive contribution which the British Ford company made to the build-up of our air forces.*[18]

Today the factory area is a modern industrial park.

The Shadow Factory system worked very well. It was also applied to the manufacture of tanks, small arms, ammunition and a host of other vitally important war supplies. The 10,000 men and 7,000 women who worked at Ford's at Trafford Park in Manchester were all proud of their achievement. It is, however, a strange fact that many aircraft enthusiasts have not heard of the Ford Merlin, yet the US Packard-built Merlin is very well known.

Packard Motor Car Company of Detroit

As the Merlin was so important to the war effort, negotiations were soon started to establish an alternative production line outside the United Kingdom. Rolls-Royce staff visited a number of North American automobile manufacturers in order to select one to build the Merlin in the United States or Canada.

In June 1940, during a discussion with US Secretary of the Treasury, Henry Morgenthau, regarding what the

Ford Motor Company might produce, Ford agreed to make 6,000 Rolls-Royce Merlin engines for Great Britain and 3,000 for the United States.

During July and August 1940, Henry Ford had a change of mind and reneged on the agreement, stating that he would manufacture only for the defence of the United States, not for Britain. Ford was not impressed by Britain's or France's chances of defeating a German onslaught and decided to keep his plants in the United States neutral. The entire deal was declared 'off'. As a result of this decision, all Merlin and other armament work of any description was cleared out of the Ford factories in Detroit. It has been said that Henry Ford turned down the chance to make the world's greatest aero engine in the United States because he thought Britain would lose the war. Thankfully, the same criteria did not apply to the Ford factory at Trafford Park in Manchester.

After the Ford debacle, members of the US Defence Advisory Commission subsequently began negotiations with other manufacturers in an effort to place the $130 million Rolls-Royce order ($2.15 billion at 2015 prices), and the Packard Motor Car Company was eventually chosen, mainly because Rolls-Royce was impressed by their engineering expertise. Agreement was reached in September 1940, and the first Packard-built engine, designated V-1650-1, ran in August 1941. By the end of hostilities in 1945, Packard had made over 55,000 Merlin engines, an extraordinary and prodigious feat of manufacturing production and output.

The first Packard V-1650s, fitted with a simple one-stage supercharger, were used in the P-40F Kittyhawk fighter. Later versions included the more advanced two-stage supercharger, greatly improving performance at high altitudes. The Packard Merlin found its most famous application in the North American P-51 Mustang fighter, where it vastly improved that aircraft's performance at altitude, transforming the Mustang into an outstanding fighter with the range and performance to escort heavy bombers over the European continent.

The most important improvement Packard incorporated into the Merlin was to adopt the Wright supercharger drive quill. This alteration meant that the two-speed two-stage supercharger section of the engine featured two separate impellers on the same shaft, usually driven through a gear train. This modification was designated the V-1650-3, and became known as the 'High Altitude' Merlin, which was used in the Mustang fighter. The ability of the supercharger to maintain a sea level atmosphere in the induction system to the cylinders at high altitudes allowed the Merlin to develop more than 1,270hp at elevations beyond 30,000 feet. The Packard factory in Detroit that produced the Merlin in such exceptional quantities is now closed down and derelict.

Along with the development of the Shadow Factory scheme, what was also needed were manufacturing engineers and managers who could produce engines in quantities never before achieved in industry. One man made that aim his daily work. That man exemplified the determination, hard work and persistence endemic throughout Rolls-Royce and British industry, both immediately before and during World War II. That man was the General Manager of the Rolls-Royce Company during World War II, Ernest Hives. His story begins our next chapter. But first a chart to show the total output of Merlin engines.

Manufacturing Organisation for Wartime Production of Merlin Engines

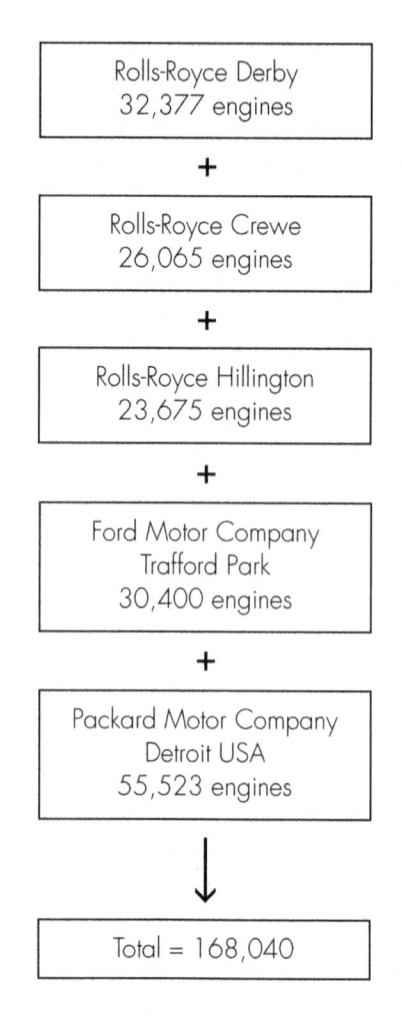

```
┌─────────────────────────┐
│    Rolls-Royce Derby    │
│     32,377 engines      │
└─────────────────────────┘
            +
┌─────────────────────────┐
│    Rolls-Royce Crewe    │
│     26,065 engines      │
└─────────────────────────┘
            +
┌─────────────────────────┐
│  Rolls-Royce Hillington │
│     23,675 engines      │
└─────────────────────────┘
            +
┌─────────────────────────┐
│   Ford Motor Company    │
│      Trafford Park      │
│     30,400 engines      │
└─────────────────────────┘
            +
┌─────────────────────────┐
│  Packard Motor Company  │
│       Detroit USA       │
│     55,523 engines      │
└─────────────────────────┘
            ↓
┌─────────────────────────┐
│    Total = 168,040      │
└─────────────────────────┘
```

Chapter 5
Inspirational Leadership

Ernest Hives

> *I can say with all sincerity that those who are best able to judge would put you among a very few men whom the country has to thank above all others for our survival in this war.*
>
> Air Chief Marshall Sir Charles Portal
> [in a letter to Ernest Hives][19]

Ernest Walter Hives was born in 1886 in Reading, Berkshire. He began his working life in a local garage, and in 1903, after he had successfully worked on the Honourable C.S. Rolls's car, he got a job working in a garage owned by Rolls. Hives graduated to become a chief test driver, and in 1913 he led the Rolls-Royce team in the Austrian Alpine Trial event.

During World War I, Rolls-Royce had designed and developed its first aero engine, the Eagle, and Hives had become, by 1916, the Head of the Experimental Department for Rolls-Royce in Derby. Hives was promoted to become Works Manager of the Rolls-Royce factories in Derby in 1936, and in 1937 he was promoted again and was elected to the Board of Directors of Rolls-Royce. Hives had a conviction and belief that war with Germany was inevitable, and accordingly he began preparing the company for a massive increase in production and output of Merlin engines during 1937. When war eventually arrived, Hives's foresight and drive proved to be of vital strategic importance. It is not unreasonable to say that without Hives's determination, the incredible output of 168,040 Merlin engines produced during the war may well not have been achieved.

Most of the piston aero engines developed by Rolls-Royce were advanced under Hives's leadership. In rough chronological order, these were the Eagle, Hawk, Falcon, Condor, Kestrel, Buzzard, Goshawk, R, Peregrine, Merlin, Exe, Vulture, Griffon, Pennine and Crecy.

Of all these engines, the Crecy was the most unusual. Liquid cooled V12 of two-stroke configuration, it used direct petrol-injection. It was developed between 1941 and 1945 and named after the battle of Crecy in France, a decisive victory for the English and Welsh forces under Edward III. Battles had been chosen by Rolls-Royce as the subject for their two-stroke engines, the Crecy being the only one ever built. This engine was intended to power the Spitfire, but never did. It was cancelled in 1945.

Perhaps the finest description of the work of Ernest Hives was given by Sir Archibald Sinclair, Secretary of State for War 1940-1945, who said in a speech at the opening ceremony for the memorial window (see Chapter 9) in Rolls-Royce's Derby factory in January 1949:

There was one frequent visitor from Rolls-Royce to the Ministry of Aircraft Production and to the commands of the RAF. He was in the confidence of the Air Staff, he was foremost amongst the many representatives of Rolls-Royce who kept in close and constant touch with the RAF and its requirements, and he seemed to us to personify the ceaseless thrusting imaginative energy, purpose and will to a victory which animated Rolls-Royce – and that was Mr Ernest Walter Hives. Long may he be spared to fertilise by his genius, and inspire by his example, the work of your hands and brains.[20]

Lord Hives wrote to the entire Rolls-Royce workforce on 20 September 1940, shortly after the victory in the Battle of Britain. In his letter he said:

"Never in the field of human conflict was so much owed by so many to so few," said the Prime Minister recently. He was referring to the RAF pilots, but his words are equally applicable to all who work in the Rolls-Royce factories which turn out engines for Spitfires, Hurricanes and Defiants.

All workers at Rolls-Royce factories have a direct share in the success of the RAF fighters. All free people are unanimous in their praise for the work of the RAF fighters – the pilots, the machines and the engines. The whole country looks forward to the day when we shall have built up our fighter air force to such a strength that no enemy machine will dare approach our shores.

Arrangements have been made by the Government to increase the supply of Merlin engines, but until these factories are in production, the whole output must be maintained and increased by the present group – Derby, Crewe, Glasgow and sub-contractors.

I am appreciative of all that has been done, and the whole country is grateful for it, but there is such a great responsibility on all Rolls-Royce workers that THERE MUST BE NO LETTING UP, EVEN AT THE EXPENSE OF PERSONAL COMFORT.

WORK UNTIL IT HURTS. *Must be our slogan!*[21]

The Vice Chief of RAF Air Staff, Sir Wilfrid Freeman, the architect behind the dramatic advances in British aircraft production before and during World War II, paid tribute to Hives's dedication in a letter to his wife:

That man Hives is the best man I have ever come across for many a year. God knows where the RAF would have been without him. He cares for nothing except the defeat of Germany and he does all his work to that end, living a life of unending labour.[22]

Hives was raised to the peerage on 7 July 1950, becoming Baron Hives of Duffield in the County of Derby. He died in 1965, aged seventy-nine.

While Hives concentrated on engine production, there were two other men during the war who, by their vision and determination, succeeded in organising the manufacturing effort to produce aircraft in extraordinary quantities. Their names were Wilfrid Freeman and Lord Beaverbrook.

Wilfrid Freeman and Lord Beaverbrook

Wilfrid Freeman, born on 18 July 1888, was one of the most important influences on the rearmament and re-equipping of the RAF in the years leading up to World War II. Freeman's military career included joining the Royal Flying Corps in 1914, where he saw active service as Officer commanding No 14 Squadron. He continued to serve in the newly formed RAF in the years between the wars, becoming Commandant of the RAF Staff College at Andover in 1933. From 1936, in his role as Air Member for Research and Development he was given the job of choosing the aircraft with which to rearm the RAF. In 1938 this remit was expanded to include the controlling of aircraft production, which he did with great distinction until November 1940. Freeman was a major influence in the RAF, not least by ordering, among others, the Mosquito, Lancaster and Halifax. He also played a significant role by his support in the development of the Merlin powered Mustang fighter. In November 1940, Freeman was moved from his role controlling aircraft production and became RAF Vice Chief of the Air Staff.

His old department, now renamed and reformed into the Ministry of Aircraft Production (MAP), was led by the newspaper magnate and Canadian, Lord Beaverbrook, who had been directly appointed by the Prime Minister, Winston Churchill, a personal friend. Beaverbrook did not stay in this position for very long, however. He resigned from his post as Minister of Aircraft Production on 30 April 1941, and there followed a series of appointments for Beaverbrook by Churchill, including the Minister of Supply and Minister of War Production. In both roles

Beaverbrook clashed with Ernest Bevin, the Minister of Labour and National Service. Ultimately, in the face of Bevin's refusal to work with him, Beaverbrook resigned, and in September 1943 he was appointed Lord Privy Seal, a position outside of the Cabinet and less likely to cause any controversy or personality clashes. This post he held until the end of the war. After the war, Beaverbrook returned to running his Daily Express newspaper empire and his charitable work. Lord Beaverbrook died in Surrey, England in 1964.

Beaverbrook's tenure as the Minister for Aircraft Production has been much debated by historians, and there is a degree of controversy over just how much he did contribute to the success of aircraft and engine output, but his aggressive style certainly provided impetus to the production of aircraft and engines at a time when it was needed. It has been argued, and is indeed a fact, that aircraft and engine production was already rising when Beaverbrook took charge, and he was fortunate to inherit a system, developed by Freeman, which was already proving itself successful in meeting the demands of war. It was Freeman himself, however, who said of Beaverbrook:

> *I wouldn't have believed it possible for anyone to go on as Beaverbrook does, day after day, with only six hours off...The odds against us at times seem overwhelming; that's one advantage in the Beaver, he does lessen the odds and override difficulties as no other politician would do, or even dare to attempt.*[23]

Following Beaverbrook's tenure at the Ministry of Aircraft Production, Freeman was recalled to lead the MAP from October 1942 to March 1945 as Chief Executive.

Major G.P. Bulman, the man who had guided R.J. Mitchell towards using Rolls-Royce to power the Supermarine seaplanes in 1929, and who was given by Freeman in October 1938 the job of overseeing all engine and propeller production at the Air Ministry, said of Freeman:

Beyond doubt, he was the most inspiring man I ever served.[24]

A further quote regarding Freeman's undoubted contribution to the war effort says:

'There was not one single monoplane in squadron service in 1936 when Wilfred Freeman accepted the task of choosing the aircraft with which to rearm the RAF. Freeman's ruthless quest for quality and exceptional organisational ability resulted in the RAF being equipped with a superb range of aircraft from the Spitfire and Lancaster to the Mosquito and the Merlin-powered Mustang.[25]

Sir Wilfrid Rhodes Freeman eventually became an Air Chief Marshall in the RAF. Post war, he worked for Courtaulds Ltd, a manufacturer of textiles and chemicals. He died in 1953.

These two men, Freeman and Beaverbrook, were immensely important to the outcome of the war. Through their persistence, levels of aircraft and aero engine production reached heights never achieved before or since.

A cover story from early 1940 in the US magazine *Time* declared:

> *Even if Britain goes down this fall* [autumn], *it will not be Lord Beaverbrook's fault...This war is a war of machines. It will be won on the assembly line.* [26]

When one considers the absolutely phenomenal fact that 168,040 Merlin engines were built mostly in a little over six years, one can only say "Amen" to that fact. The war of production, of output, was won on the assembly line.

Chapter 6
The Architects of Victory

What follows in this chapter is a brief biographical history of significant individuals involved in the design, development and production of the Merlin engine and the designers of the key aircraft powered by it in World War II. Again through reasons of space I have included just a few – there are many others not mentioned here whose contribution to Allied success in World War II should not be ignored. This whole story really is about a remarkable team effort, even though the team involved never got together as one body, but worked over time and distance towards a common aim: victory.

As we have read previously, one of the more important development moments in the Merlin's origins, if not the most important, was the winning of the Schneider Trophy races in 1931 by the Supermarine S.6B seaplane, and the use of the Rolls-Royce R engine in that successful programme. The 1931 races and the accelerated development in engine design they produced would not have taken place but for one wonderfully eccentric, patriotic and generous individual: Lucy, Lady Houston. Her story deserves to stand alongside those of other great Britons.

Lucy, Lady Houston

Lady Houston certainly had a colourful and interesting life. Born in 1857 in Lambeth, London, Fanny Lucy Radmall, the daughter of a warehouseman, was the ninth of ten children. At the age of sixteen, determined to gain entry into show business, she presented herself at the front office of the Drury Lane Theatre in London and asked to speak to the recently appointed manager, one Augustus Harris. She was courteously told to go away as there were no vacancies available at that time. Being nothing if not determined, she took no notice of the rejection. Instead she resorted to the no-nonsense approach which was going to become her trademark in the future when she was telling heads of state and other important people what she thought they should be doing – she brushed aside the stage door staff that stood between her and her goal and marched into the head man's office. Harris was, by many accounts, charmed by the beautiful brown-eyed teenager and employed her on the spot.

Lucy became a professional dancer and chorus girl at the theatre and acquired the sobriquet 'Poppy'. She attracted the attention of many suitors and soon ran away to Paris with a wealthy married man, Frederick Gretton, whose family owned the Bass brewing empire. Upon his death in 1882, Gretton bequeathed her £6,000 per year for life in his will. They had never married.

After Gretton's death, Lucy, by then aged twenty-five, returned to London and bought a house in one of the most fashionable parts of the capital: Portland Place, Marylebone. She still had enough money left to employ a butler, a lady's maid and a coachman. The years she had spent in Paris had been the making of Lucy. Not only had she learned how to use and manipulate the many men who sought her favour, but she had realised that there was more to life than just buying whatever one wanted. Lucy was always one of the best dressed woman at any function and had many of the young men of London at her feet. In later life she claimed that Winston Churchill was one of her early conquests, and if she had felt so disposed, she could have married him. Well, no one will ever know the truth of that, though it is said that the great man retained a reluctant admiration for her for the rest of his life.

Lucy wanted to marry again, preferably someone with a title, and she found what she wanted in a young, vigorous cavalry officer. His name was Theodore Brinckman. At twenty-one, he was five years younger than Lucy when they married in 1883, and he was by then Lt Colonel Sir Theodore Brinckman. Once married, however, it appears he acted his new role only part-time. For him, weekends were for carousing and womanising, not for playing the part of a devoted husband. When Lucy found out he had a second 'wife' in an apartment in London, she went to see his lover and confronted her. Realising that her marriage was now just a sham, Lucy decided that divorce from Brinckman was the best option. Their divorce was finalised in 1895 after twelve years of marriage, the later years of which were marked by a long separation. This marriage, however, had given Lucy a taste for being numbered among the British aristocracy, something she loved and did not want to relinquish. All she needed was a more senior peer of the realm – a Lord, perhaps.

She subsequently met Lord Byron, the ninth person to

hold the title. They married in 1901, and the ninth Lady Byron became London's queen of entertaining. If you didn't get an invitation to dinner at Byron Cottage, Lucy's home in Hampstead, you were not moving in the right circles. The greatest celebrities of the day often joined her for dinner and parties. She listened to the likes of author Rudyard Kipling, and Cecil Rhodes, the businessman and politician after whom Rhodesia was named, discussing the British Empire, and she was completely captivated by the tales they told and the vistas they opened. Britannia ruled the waves, and little Lucy Radmall from Camberwell was in a position to take her place among the rulers. It was during this marriage that she was an active suffragette, campaigning vigorously for votes for women.

The ninth Lord Byron died in 1917, and it can hardly be said that his death had a great effect on Lucy. The pair had been living in separate rooms for a number of years and were not the most loving of married couples. Since the start of the war, Lucy had become a frequent visitor to the many hospitals where the wounded from the Western Front were sent to recuperate, and she was impressed and astounded by the work ethic and unrelenting devotion to duty of the nurses. To Lucy, it appeared that the nurses never ceased working and caring for their patients, a situation that often led them to a state of near physical and mental breakdown. Because of this, Lucy decided to finance and set-up a rest home for nurses in the London Borough of Hampstead. When Rudyard Kipling heard about this rest home he became one of its most passionate supporters, and even King George V was impressed with the work Lucy was doing. In 1917, for championing nurses and setting up the home, Lucy was awarded the honour of Dame of the British Empire, only the fifth woman ever to be made a DBE.

Lucy married for the third time in 1924. Her husband, Sir Robert Houston, the Member of Parliament for West Toxteth in Liverpool, was famously described in the Oxford Dictionary of National Biography as "...a hard, ruthless, unpleasant bachelor".[27] After their wedding, they lived on the island of Jersey as tax exiles. When Sir Robert made his original will, he left Lucy £1 million. He showed it to her before he left on a business trip, hoping to impress his wife, but he returned to find the will torn in half, lying on his desk. When he challenged Lady Houston about it, she is reported to have said, "If that's all I'm worth to you, I suggest you leave me out and spread your money among those more deserving." The following day he showed her his new will. She would receive the bulk of his fortune – an enormous £5,500,000.

Sir Robert and Lucy were both of strong personalities and forthright views. She was as opposed to the drift towards Socialism, which in their opinion equated with Communism, as she was to the disgraceful way that the heroes of World War I were treated after demobilisation at the end of the war. Many ex-servicemen suffered as unemployment rose rapidly and the ambitious wartime programme of reconstruction and job creation was postponed during the economic slump of 1921. All this in the 'land fit for heroes' as promised by the Lloyd George Government.

Sir Robert died in April 1926 while on board his yacht, the *Liberty*. In his will he did leave Lady Houston, as promised, over £5.5 million (£300 million at 2015 prices).

Lady Houston's Gift

As 1931 dawned, the Supermarine Company and Reginald Mitchell, their Chief Designer, were earnestly hoping the Government of the day would put up the money to enable the Schneider Trophy to be captured for the third time. However, on 15 January 1931, the Air Ministry refused a last minute request by the Royal Aero Club for funds to allow an entry to be made in the Schneider Trophy races. The Air Ministry further prohibited the use of aircraft that had competed in the 1929 races and banned the RAF pilots of the High Speed Flight from taking part. Furthermore they said they would not police the race course, which was to be over the busy shipping lanes of the Solent.

The Royal Aero Club sent a statement to the Cabinet on 22 January, offering to raise £100,000 if the Government would rescind the Air Ministry decrees on pilots, planes and policing. The Labour Government, led by the Prime Minister, Ramsey Macdonald, decided not to finance the attempt. There may well have been good reasons for this, including the onset of a worldwide depression caused by the Wall Street crash of 1929. Nevertheless, Lucy, Lady Houston, felt that Britain must on no account be left out of the contest. She contacted the Prime Minister and stated that she would guarantee £100,000 if necessary towards the cost of competing. This gift, and the ensuing publicity it generated, left the Government with little alternative but to reverse their previous decision. Many newspapers of the day backed the Conservative Party opposition in putting pressure on the Labour Government, and Lady Houston's gift was an ideal opportunity for them to turn up the heat. Lady Houston herself was no fan of the Labour administration. She declared:

Every true Briton would rather sell his last shirt than admit that England could not afford to defend herself.[28]

Following Lady Houston's gift, there were only nine months left to prepare for the competition. R.J. Mitchell, in the short time left available, could only update the existing airframe of the competing seaplanes for Supermarine, while Rolls-Royce increased the power of the R engine by further modifications from 1,900hp to 2,300hp.

Lady Houston's gift of £100,000 provided the impetus for the development of aero engine technology that would ultimately be vital in the Battle of Britain, leading as it did to the world-beating Merlin engine. The endowment by Lady Houston allowed Supermarine to compete and win for the third consecutive time in the 1931 race.

Lucy Houston's gift to the British competitor, Supermarine, was without precedent. Air Chief Marshal Sir Michael Graydon, president of the Battle of Britain Memorial Trust since 1999 and former Chief of the Air Staff, said of Lady Houston's gift:

[The donation] *was the incentive to develop a high-speed aeroplane. We were just about able to prepare in time for Hitler's air armada, but we got away with it only by a gnat's eyebrow.*[29]

Later Life

Lady Houston continued in her eccentric manner to embarrass the administration. In 1932 she offered £200,000 to the Government to strengthen the British Army and Navy. She was turned down. This refusal prompted her to hang a large electric sign in the rigging of her yacht, the *Liberty*, saying 'Down with Macdonald, the traitor' (Ramsey Macdonald was Prime Minister at the time). With this she sailed around Great Britain. Subsequently, in a telegram to the Prime Minister, she wrote:

I alone have dared to point out the dire need for air defence of London. You have muzzled others who have deplored this shameful neglect. You have treated my patriotic gesture with a contempt such as no other Government would have been guilty of toward a patriot.[30]

In 1933 she financed the Houston Mount Everest flight expedition, in which aircraft flew over the summit of Everest for the first time, and in 1934, Lady Houston sent a cable to the winners of the England to Melbourne air race, T. Campbell-Black and C. Scott:

Your achievement has thrilled me through, oh brave men of my heart...If this does not make the Government sit up, nothing will...Sleep well and feel proud of yourselves, as we all are...Rule Britannia. God bless you both.[31]

Lady Houston died of a heart attack on 29 December 1936, aged seventy-nine, and is buried in St Marylebone

Cemetery in North London (now known as East Finchley Cemetery). Her headstone describes her as 'One of England's greatest patriots'. Many believe she deserves greater recognition. Some say that a statue of her could and should permanently adorn the fourth plinth at Trafalgar Square. The statue could show Lady Houston looking as she did when she was the toast of Edwardian London, with perhaps a suitable inscription: *Lucy Huston – The First of the Few.*[32]

Air Chief Marshall Sir Michael Graydon does not disagree. To criticisms that she was at least a little mad, he is quoted as saying: "Well, thank God for her madness!"[33] Indeed, Marshall of the RAF, Lord Trenchard, once suggested that there should be three monuments erected on the White Cliffs of Dover. They should show Winston Churchill, an Unnamed Airman and Lady Houston. It could be argued that a national memorial to Lady Houston is long overdue.

Lady Houston's patriotic gift secured the Schneider Trophy and enabled the development work necessary to be undertaken on improving aero engine performance. Great Britain had the world-beating R engine. Now it needed engineers who could carry the progress made in engine performance forward into what would become the Merlin. Those engineers mostly came from a company that had been set up by two men, one of whom, Charles Rolls, was now dead, and the other, Sir Henry Royce, was in semi-retirement in 1931.

Henry Royce

Frederick Henry Royce was born in Alwalton, near Peterborough, in 1863, the youngest of five children. His family ran a flour mill, but their business eventually failed and they moved to London, where in 1872 Royce's father died. Thereafter, from the age of nine Royce had to go to work to supplement the family income. He delivered newspapers and telegrams, until in 1878 he started work as an apprentice with the Great Northern Railway Company in Peterborough. In this he was helped with some financial assistance from an aunt. His aunt's money ran out after three years. Having received only one year of formal schooling, Royce returned to London after a short period at a tool making company in Leeds and was employed by the Electric Light and Power Company, moving to their Liverpool office in 1882 where he worked on street and theatre lighting.

In 1884, at the age of twenty-one Royce set up a company that made electrical fittings in Hulme, Manchester. This business, named F.H. Royce and Company, was a partnership with a friend of Royce's, Ernest Claremont. The business prospered, and in 1894 they expanded into the manufacture of electric dynamos and electric cranes. In 1899 a further factory opened in Trafford Park, Manchester following a public share flotation.

With his absolute fascination for engineering and all things mechanical, Royce became interested in the new-fangled device for getting from A to B: the motor car. Accordingly, he bought a 1901 De Dion and a Decauville, both French automobiles. Detailed examination of these two cars taught Royce a great deal about automotive engineering, and ultimately he built his own car and engine in his workshops in 1904. Aluminium and brass castings for the new engine were made in Manchester, while iron castings were made at the factory at Trafford Park. Royce recognised that better performance and absence of vibration would best be achieved by lightness in all mechanical components, and he pioneered this area of engine development. Encouraged, Royce made two more cars, one of which was given to Ernest Claremont, and the other sold to one of the Company's directors, Henry Edmunds. Edmunds was a friend of Charles Rolls, who at that time had a car showroom in London selling imported cars. He showed Rolls his new car, and subsequently arranged the historic meeting between Royce and Rolls at the Midland hotel in Manchester on 4 May 1904. Rolls was impressed with the two-cylinder Royce 10, and in a subsequent agreement on 23 December 1904, he contracted to take all the cars Royce could make. It was agreed that these would be badged and identified as manufactured by Rolls-Royce.

The very first Rolls-Royce car, the 1904 Rolls-Royce 10hp Two Seater number 20154, was unveiled at the Paris Salon Motor Show in 1904. It is still running today, and was last sold by the auctioneers Bonhams in 2007 for £3,500,000.[34]

Sir Henry Royce lived by the motto "Whatever is rightly done, however humble, is noble".[35] Indeed, this maxim was engraved above the fireplace at his home. He was awarded the OBE in 1918, and made a Baronet in 1930 for services to British aviation.

The name forever linked with Sir Henry Royce is, of course, that of Charles Rolls.

Charles Rolls

Charles Stewart Rolls, was born into a privileged background on 27 August 1877, the youngest son of Lord and Lady Llangattock, whose family seat was in Monmouthshire.

Displaying an early engineering bent, and a fascination with electricity in particular, young Rolls went up to Trinity College, Cambridge in 1895 to study electrical and mechanical engineering. As a student at Trinity, he must have been the envy of his fellow students when he returned from a Paris trip in October 1896, the owner of a 3 3/4hp Peugeot Phaeton, the first ever car used at Cambridge. Thus began his fascination with the motor car, and Rolls became a committee member of the newly formed Automobile Club of Great Britain and Ireland. He actively participated in motor tours and races, and in 1903 he briefly held the unofficial World Land Speed Record when he drove an 80hp Mors car at 83mph along a course in the Duke of Portland's Clipstone Park. Rolls fully recognised the potential of the motor car, passionately proclaiming its virtues at every available opportunity.

Following the famous meeting between Rolls and Royce in Manchester in 1904, the partnership between them was formalised in 1906, and Rolls-Royce Limited was created. Royce was appointed Chief Engineer and Works Director and would provide technical expertise. This would complement the business acumen and financial backing of Charles Rolls. The company was soon winning awards for the reliability of their motor cars.

Rolls himself was certainly an adventurer. On 2 June 1910, he became the first man to make a non-stop double crossing of the English Channel by plane, taking 95 minutes. For this feat, which included the first east-bound aerial crossing of the English Channel, he was subsequently awarded the Gold Medal of the Royal Aero Club.

The partnership between Rolls and Royce came to an

end when Charles Rolls, aged thirty-two, was tragically killed in a plane crash on 12 July 1910 at Hengistbury airfield, near Bournemouth, when the tail of his Wright Flyer aircraft broke off during a flying display. He was the first Briton to be killed in an accident with a powered aircraft, and the eleventh internationally to suffer this fate.

After Rolls's death, Royce continued to build up the Rolls-Royce Company, and turned it into a very successful business. Today it has one of the most recognisable trademarks in the world, and is a byword for quality in all things.

During his tenure as the General Manager of Rolls-Royce, Sir Henry had set in motion the design momentum that would lead to the Merlin engine. His company was very fortunate in having a number of quite brilliant design engineers who were able to realise the full potential of the legacy left by the R engine and the PV12. One of those design engineers was Cyril Lovesey.

Cyril Lovesey

Cyril Lovesey, having obtained a BSc from Bristol University, joined the Rolls-Royce experimental department in 1923 at the age of twenty-four. He was the company representative for support on the R engine during the Schneider Trophy races in 1929 and 1931. A proponent of flight testing, Lovesey established the testing centre at RAF Hucknall where he was Flight Development Engineer. In the mid-1930s, Lovesey began working on the Merlin engine, and in 1940 he became Rolls-Royce Chief Experimental Engineer and was in charge of the Merlin development programme. In this role he was responsible for the mechanical integrity of the Merlin through its many, many modifications and development. It was generally accepted at that time that Cyril Lovesey was one of the finest development engineers in the world.

During the war, engineers from all over England often met to pool their knowledge and ideas in the true wartime spirit of cooperation. During these meetings it became apparent that many non-Rolls-Royce engineers believed that a turbocharged Merlin engine was the best way forward in its development. This situation escalated to the point that Government pressure was applied to Rolls-Royce to look at the possibility of a turbocharged Merlin. In exasperation, Ernest Hives, the General Manager, sent Lovesey to meet these engineers and give them a talk on the Merlin. Though no record of the meeting was made, Lovesey said words to this effect to the assembled engineers:

> *For some time now, as you all should know, my team and I have been developing the 'rear facing' exhaust stubs on the Merlin engine. I have calculated that the jet propulsion effect of the stubs on the Spitfire at full speed will produce the equivalent of close on 100 Brake horsepower. I am therefore loath to sacrifice this by fitting an exhaust turbo supercharger into the system which, besides relieving me of this power, would threaten aerodynamic integrity and give us problems with the airframe. It would also involve my staff in a plumbing problem almost as complex as that of improving the toilet facilities on Plymouth*

Naval Docks. (The implication being, naval docks were blessed with row upon row of toilets for the sailors, most of which were often out of order due to 'plumbing problems').[36]

After this talk there was no more mention of a Merlin turbocharger.

Lovesey was a great engineer. He continued working for Rolls-Royce until 1964, when he retired. In 1971 he was recalled, along with Sir Stanley Hooker and Arthur Rubbra, to save Rolls-Royce following its bankruptcy due to faults with the RB211 gas turbine.

Arthur Rubbra

Arthur Alexander Cecil Rubbra was born in Northampton in 1903, the second son of a Northampton watchmaker. His rare surname is a form of Ruborough, a village near Bloomfield in Somerset where his ancestors may have originated. Because their parents were not well-off, both Arthur and his brother, Edmund, the distinguished composer, had to make their way in life through hard work and their exceptional talents. Arthur attended Northampton Grammar School. He won a scholarship to study at Bristol University, graduating in 1925 with a BSc in engineering. As a young man, Rubbra was very interested in steam engines and loved to watch the steam trains of the day.

Rubbra's first job was with the Armstrong-Siddeley Company, where he worked until 1925. Then he joined Rolls-Royce in Derby as an Assistant Tester in the engine testing department. He worked on the Eagle, Kestrel and Buzzard aero engines, and in 1927 he was promoted to become a Design Engineer. This promotion led to further work on the Buzzard and Kestrel, and on the new R engine for the Schneider Trophy races. In 1934 he was appointed Assistant Chief Designer, and in 1940 again promoted to Chief Designer, Aero Engines. His design work continued with the development of the Merlin and the Griffon.

Rubbra could impart much information with few words, and he was renowned as a calming, tranquil influence at many senior design meetings, where, according to colleagues, he was always two jumps ahead. Because of the speed of his comprehension, there was no need to explain anything twice; Rubbra was quick to grasp whatever new idea was debated.

Rubbra was awarded a CBE in 1961. In 1966, because of poor health, Rubbra gave up his post as Technical Director of Rolls-Royce, although he continued to work as Chief Technical Advisor. He retired later in 1966, having become the Vice Chairman of Rolls-Royce. In the words of Lionel Haworth, Arthur Rubbra was "A gentle giant"[37] who has not had his fair share of praise for all his achievements. In 1969 Rubbra was awarded the Royal Aeronautical Society's highest honour, the Gold Medal, for outstanding contributions over many years in the whole field of aircraft propulsion. He worked with Stanley Hooker again in 1971 when they were asked to help with the technical problems of the RB211 turbofan engine which had caused the financial collapse of Rolls-Royce. They, and their team of engineers, succeeded in obtaining RB211 engine certification in April 1972, following which the engine became an outstanding success. A.A. Rubbra died in November 1982 aged seventy-nine.

Rubbra was a great engine designer, of that there is no doubt. While at Rolls-Royce, Rubbra would have collaborated with another great designer; a man who titled his autobiography, probably a little tongue in cheek, *Not Much of an Engineer* – Dr Stanley Hooker.

Stanley Hooker

Stanley Hooker was born on 30 September 1907 in Sheerness, Kent. He was educated at Borden Grammar School in Sittingbourne. Hooker won a scholarship to Imperial College, London, where he studied Mathematics. Subsequently, in 1935 Hooker received his DPhil in aerodynamics from Brasenose College, Oxford.

In 1937, while working at the Scientific Research Department of the Admiralty, Hooker applied for a job with Rolls-Royce. He was interviewed by E.W. Hives and given a role starting in January 1938. What exactly that role was is unclear, as it was even to Hooker at that time. Hooker was allowed to study anything that took his fancy. In his own words, he was left to himself in an office where he sat and read and smoked all day. Soon tiring of this inactivity, he left his office on one occasion to undertake a tour of the adjacent offices. Spying a man sitting in an office corner with a bunch of papers, Hooker asked what he was doing. "Testing superchargers" came the reply. Hooker was intrigued, took the papers from the man (with his permission) and, after studying them, wrote an analysis of the design and suggested improvements. The analysis was forwarded to the Chief Designer, and Hooker returned to his office and sat and read and smoked, thinking that nothing would happen.

Nothing did happen until about two weeks later, when Hooker's office door burst open and in came the Chief Experimental Engineer, Mr Eller.

"Did you write this?" he asked, waving Hooker's analysis.

"Yes, I did," responded Hooker.

"Right, you are now in charge of Supercharger development."[38]

So began the improvement of superchargers used on the Merlin, and that led to the immensely crucial advances in its capability and efficiency.

At the time Hooker joined Rolls-Royce, many feared that Europe would quite soon be engulfed in war. The Nazi Luftwaffe had shown its capabilities at Guernica in the Basque region of Spain during the Spanish Civil War, where the Condor Legion decimated the town through bombing. The Condor Legion was a force of volunteers from the German Air Force and the German Army, who fought on the side of the Nationalists, under General Franco, during the Spanish Civil War from 1936 to 1939. The Condor Legion developed and used their bombing offensive as a terror weapon, and these methods of bombing were subsequently used by the Nazi Luftwaffe during World War II.

The Government of Britain was now becoming desperate to get RAF Fighter Command up to a suitable strength, so that if war came, Britain would stand some chance of repelling any invasion threat. Despite the belief that Merlin-powered Hurricanes and Spitfires were a match for the German Me109, the early Merlin engine had one big weakness in that its supercharger could not produce sufficient power to stay ahead of the opposition. This weakness could have been fatal.

Stanley Hooker took a mathematical approach to the supercharger problem, and his initial changes allowed a thirty per cent increase in power and propelled the aircraft 22mph faster. Hooker's subsequent improvements to the supercharger, described very well by himself in the filmed interview mentioned above, kept the Spitfire well ahead of the Nazi fighters.

The Mark of engine that incorporated Hooker's improved supercharger was the Merlin XX. The increased power was to be of great importance to the Spitfire and Hurricane during the Battle of Britain, where a great deal of aerial combat was below 6,000 feet – an altitude at which the previous supercharger had not worked very well. The supercharger of the Merlin XX was still weaker at altitudes above 6,000ft, however, and to overcome this weakness in order to attack high flying Luftwaffe bombers, Hooker incorporated a further modification that used two superchargers in a series. This modified engine became the Merlin 61, which was fitted to the Spitfire Mark IX. Hooker was a modest man who underplayed his achievements. As previously mentioned, he titled his autobiography *Not Much of an Engineer* – something that had been said to him by Lord Hives during his interview for a job at Rolls-Royce in 1938. Throughout his career, Hooker was a 'hands on' engineer in that he kept very close to the work and development of any new engine for which he was responsible.

One of his great achievements for Rolls-Royce was the Dart engine, a successful turbo-prop that powered the Vickers Viscount civil passenger airliner and eleven other aircraft. While in Rolls-Royce Bristol, he worked closely with Lionel Haworth in designing and developing the engines that powered the Harrier fighter, the Concorde civil airliner and the Tornado fighter.

In 1970, Hooker retired from Rolls-Royce, saddened, according to some, that he had never achieved his aim of becoming Director of Engine Development. Following the financial collapse of Rolls-Royce in 1971, Hooker was persuaded to return to the Company as Technical Director to lead a team of other retirees to solve the problems with the RB211, for which he was knighted in 1974. He retired again in 1978, and died on 24 May 1984.

In Sir Stanley Hooker's own words from a filmed interview in 1981:

> *I think my greatest success was the two stage supercharger for the Merlin, that I did do myself. I proposed it, I specified it, did all the calculations. We* [Rolls-Royce] *made it and it worked first pot. It put 70mph on the speed of the Spitfire and 10,000 ft. on its fighting altitude. The pilot, and later Air Vice Marshall, who flew the first one often tells me how he soared up past a FW190 and he could see the German pilot follow him up with a look of utter astonishment on his face...We were all obsessed that we were going to keep the Merlin ahead of the Germans. And we did too.*[39]

Just how important Hooker's work was to the eventual outcome of the battle is concisely stated by Lionel Haworth:

By these means the RAF was able to keep ahead of the Germans. If it had not been for that work [Stanley Hooker's supercharger technology] *I don't think we would have won the Battle of Britain.*[40]

Bill Bedford, a British test pilot and pioneer of the development of vertical take-off aircraft, gave a talk in Christie's Auction Room, South Kensington, London in the late 1980s. He had been the original test pilot for the Harrier fighter at Dunsfold Aerodrome, and his talk was about the various fighters he had flown, many of which had been powered by Hooker's engines. On the screen behind him, towards the end of his talk, he showed a picture of Hooker, and said:

I'll have to think about this a bit, but if I was asked who was Britain's greatest ever engineer, I'd have to decide between Brunel and Sir Stanley Hooker, but I'd probably go for Sir Stanley.[41]

Bedford was the first pilot to fly the V/STOL (Vertical/Short Take Off and Landing) Kestrel and Harrier jump jet, and the first pilot to fly a Harrier aircraft from a ship off HMS *Ark Royal* in 1963.

Uniquely in this chapter, there is one lady engineer. Her contribution to victory in the ongoing struggle between the Allies and the Axis air forces was very important indeed. Her name was Beatrice Shilling.

Beatrice Shilling

Beatrice Shilling was born in Waterlooville, Hampshire, in 1909, the daughter of a butcher. Her first job after leaving school was working for an electrical engineering company, installing wiring and generators. Her employer encouraged her to take a degree in Engineering at Manchester University. This she did, and eventually graduated with an MSc in Mechanical Engineering. She was then recruited as a scientific officer by the Royal Aircraft Establishment (RAE) at Farnborough, where she worked for most of her life. Affectionately known as 'Tilly', Shilling is best remembered for solving the problem with the Merlin SU Carburettor.

During the Battle of France and Battle of Britain in 1940, it became apparent that the Hurricane and Spitfire had a serious problem with their carburettors while manoeuvring in combat. The negative G force created by suddenly lowering the nose of the aircraft resulted in the engine being flooded with excess fuel, causing it to lose power or shut down completely. German fighters used fuel-injection engines and did not have this problem, so during combat they could evade RAF fighters by flying negative G manoeuvres that could not be easily followed by the pursuing RAF pilots. A German pilot with a Spitfire or Hurricane fighter on his tail could simply pull negative G by nosing into a dive, and the Spitfire would fall behind until the engine picked up. This took only a matter of a second or two, but that second was all the German needed. Some RAF pilots did overcome this disadvantage by doing a half-roll before following into a dive. This meant that the force of gravity acted in the opposite direction and the Merlin would then not choke on too much fuel. However, the time taken to do the half-roll often put the RAF pilot at a disadvantage.

Shilling's solution was to add a small metal disc with a hole, similar to a washer, into the carburettor, which restricted the fuel during a negative G dive of the Spitfire. Although not a complete solution, this allowed RAF pilots to perform quick negative G manoeuvres without loss of engine power. By March 1941, Miss Shilling had led a small team on a tour of RAF fighter bases throughout the country, installing the device in their Merlin engines. The RAE Restrictor, to give it the correct title, was immensely popular with pilots, who affectionately named it 'Miss Shilling's Orifice', or more simply the 'Tilly Orifice'. It continued in use as a stop-gap until the introduction of the pressure carburettor in 1943.

After the war, Shilling worked on a variety of projects, including the Blue Streak Missile and the effects of a wet runway on braking. She was described by a fellow scientist as 'a flaming pathfinder of women's liberation', and she always vetoed any suggestion that as a woman she might be inferior to a man in technical and scientific expertise.

As a woman of original view and strong personality, with a brusque manner and a sometimes adversarial attitude to bureaucracy, she had an uneasy relationship with management. Shilling worked for the RAE until the late 1960s, but never achieved high rank within the organisation.

Shilling loved racing motorcycles, and was awarded the Gold Star for lapping the Brooklands track at over 100mph on her Norton M30 motorbike. After World War II, she raced cars with her husband, George Naylor, an RAF bomber pilot who reached the rank of Wing Commander. It is said that Shilling refused to marry Naylor until he had proved himself by also being awarded the Brooklands Gold Star for lapping Brooklands at over 100 mph. He duly did!

In common with Reginald Mitchell, Beatrice 'Tilly' Shilling achieved that great honour for a British commoner of having a public house named after her. The Tilly Shilling Pub opened in Farnborough in 2011, and is dedicated to the RAE engineer who made an important contribution to the success of the Merlin engine during the struggle for air supremacy.

In our next chapter we briefly discuss five aircraft designers who were great engineers in their respective areas of expertise, and who all made an immense contribution to the realisation of the Merlin as the Saviour of the Free World. The first of the five is Reginald Mitchell, the design genius who will forever be associated with the Spitfire.

Chapter 7
Aircraft Designers Who Used the Merlin

Reginald Mitchell

R.J. Mitchell is probably one of Britain's, and the world's, most famous design engineers. This is largely due to one aeroplane, the Spitfire, though Mitchell was also involved in many other design projects with Supermarine and Vickers. The enduring quality of the Spitfire's design is proved when any competition that asks the general public the question "What is the greatest design of all time?" usually results in the Spitfire winning first place, with the London Tube map a close second.

Mitchell was born in 1895 in Newcastle-under-Lyme. He joined the Supermarine Aviation Works in Southampton in 1917, advanced quickly within the company and was appointed Chief Designer in 1919, Chief Engineer in 1920 and Technical Director in 1927. Indeed, he was so highly thought of in the industry that when the Vickers Company took over Supermarine in 1928, it was on the condition that Mitchell remain as a designer for the next five years. Between 1920 and 1936, Mitchell designed twenty-four aircraft, including seaplanes, fighters, bombers, flying boats and light aircraft.

Mitchell is remembered and renowned for designing the Supermarine S.6B seaplane. He was determined to perfect the design of the racing seaplane, and to that end he designed the Supermarine S.4, S.5 and S.6. The culmination of his dream was the S.6B that won the Schneider Trophy outright for Great Britain in 1931.

When starting work on the S.6 for the 1929 Schneider Trophy Races, Mitchell made a very important decision that was to affect all his future designs. Until this time, all of his Schneider Trophy aircraft had been fitted with Napier Lion engines, which had proved reliable over the years. After 1927, it was considered that the Napier engine had been developed to its limit, but Mitchell found it very hard to make a change. Unable to make up his mind (Sir Henry Royce said of Mitchell, "He is slow to decide but quick to act"), he sought the advice of Major G.P. Bulman, the Air Ministry official responsible for the development of aero engines. The only possible alternative to Napier was an engine from Rolls-Royce, and Mitchell knew very little about the firm then. According to some accounts, the conversation with Bulman went something like this:

Mitchell: "What do you think of these chaps at Derby?"

Bulman, who knew Rolls-Royce well, said he had a hunch they could do it given a chance.

"Right," replied Mitchell, "that's settled it!"

To make quite certain that he was doing the right thing, Mitchell met and discussed the question of an aero engine with Sir Henry Royce, and so began a close association with the firm which was to last for the rest of his life. Royce provided assurances to Mitchell that an engine of at least 1,500hp would be supplied to Supermarine in time for the races. The engine would have the further development potential of up to 1,900hp, with, importantly, no increase in frontal area, which would have made a fighter aircraft less aerodynamic and consequently slower. Rolls-Royce had six months to produce the power unit. Infused with the competitive spirit and desire to see Great Britain win the Schneider Trophy, the company began developing the engine which was eventually to have a very important part to play in the pedigree of the world beating Merlin. The on-time result delivered to Supermarine was a fully tested R engine. One can only wonder if Major Bulman ever learned just how important to the free world his comments to Mitchell were.

Mitchell was authorised by Supermarine in 1933 to begin a new design, an all-metal monoplane. This was initially a private venture by Supermarine, but the RAF soon became interested and the Air Ministry put up the money to develop a prototype.

Mitchell's genius lay in bringing many different types of recent engineering innovations together in one package. The famous elliptical wings of the Spitfire were originally designed by a Canadian aerodynamicist, Beverley Shenstone. The under-wing radiators had been first designed by the Royal Aeronautical Establishment at Farnborough. The monocoque construction, a structural approach to engineering that supports loads through an object's external skin, similar to an egg shell, was first developed in the United States. Mitchell brought all these together, and contributed his experience gained from high speed flight in the Schneider Trophy races. The first Spitfire flew on 5 March 1936 from Eastleigh in Hampshire, reaching 349mph. The RAF then ordered 310 of the aircraft.

Rumour had it that the German Heinkel He70 was an influence on the Supermarine Spitfire's distinctive elliptical wing, though this is very unlikely. The He70, used by Rolls-Royce to test the PV12 in exchange for Kestrel engines, did not arrive in Britain until three weeks after the Spitfire's maiden flight. Beverly Shenstone, Mitchell's aerodynamic advisor, refuted the idea that the Spitfire wing was copied from the He70. In Alfred Price's *Spitfire – A Documentary History*, Shenstone is quoted as saying:

> *It has been suggested that we at Supermarine had cribbed the wing shape from that of the He70 transport. This was not so. The elliptical wing had been used on other aircraft and its advantages were well known. Our wing was much thinner than that of the Heinkel and had a quite different section. In any case it would have been simply asking for trouble to have copied a wing shape from an aircraft designed for an entirely different purpose.*[42]

Shenstone said that the He70's influence on the Spitfire design was limited to use as a benchmark for aerodynamic smoothness. After Mitchell saw the He70 perform with the Kestrel engine fitted, he wrote a letter to Ernst Heinkel:

> *We…were particularly impressed, since we have been unable to achieve such smooth lines in the aircraft that we entered for the Schneider Trophy races…In addition to this, we recently investigated the effect that installing certain new British fighter engines would have on the He70. We were dismayed to find that your new aircraft, despite its larger measurements, is appreciably faster than our fighters. It is indeed a triumph.*[43]

Mitchell was not enamoured of the name 'Spitfire'. He described it as "Just the sort of bloody silly name they would choose."[44] One of the other names being considered for the aircraft by Supermarine was the 'Shrew' – a name that doesn't quite have the same resonance as the one that was settled upon.

Ever a realist, Mitchell gave some advice to Vickers Supermarine test pilot Jeffrey Quill during Spitfire prototype trials concerning his engineering staff. It shows how he felt about his role as a designer:

> *If anybody ever tells you anything about an aeroplane which is so bloody complicated you can't understand it, take it from me: it's all balls.*[45]

Dogged by ill-health, Mitchell was treated for cancer in 1933. Despite this, he continued to work on the Spitfire and on designs for a four-engine bomber. Mitchell also took flying lessons, gaining his pilot's licence in 1934. He was diagnosed with cancer again in 1936 and subsequently gave up work. By spring 1937, a year after the Spitfire's first flight, Mitchell was given only a few months to live. He died on 11 June 1937 at the early age of forty-two, and was succeeded by Joseph Smith as Supermarine's Chief Designer. Smith had worked on the Spitfire project since the outset, and he now oversaw the evolution of the Spitfire's design throughout World War II when it served in every area of combat. As historian J.D. Scott noted:

> *If Mitchell was born to design the Spitfire, Joseph Smith was born to defend and develop it.*[46]

Mitchell also has a public house named after him. The Reginald Mitchell pub is in Hanley, near Stoke-on-Trent.

Roy Chadwick

Roy Chadwick was born at Farnworth in Widnes in 1893, the son of a Mechanical Engineer. He attended St Clements Church School in Urmston and studied at night school from 1907 to 1911 at the Manchester Municipal College of Technology. It was during this time that he worked as a draughtsman for British Westinghouse in Manchester. Immensely talented as a designer, Chadwick began working for the Avro Company as the personal assistant of Alliot Verdon-Roe, the company's chief. Under Verdon-Roe's direction, Chadwick designed a succession of groundbreaking aircraft, including in 1912 the world's first monoplane (single-winged) aircraft with an enclosed cabin for the pilot, the Avro F. All this by the age of nineteen.

Chadwick became Avro's chief designer. Over the next thirty years, Chadwick's design output of pioneering aircraft was prodigious, and included Avro D, Avro F,

Avro 500, Avro 501, Avro 502, Avro 504, Avro Pike, Avro Baby, Avro Aldershot, Avro Avenger, Avro Commodore, Avro Anson, Avro Manchester, Avro Lancaster, Avro York and the Avro Shackleton. Chadwick also oversaw the initial designs of the Avro Vulcan bomber in 1946. In 1943, Chadwick was appointed a Commander of the Order of the British Empire in the King's Birthday Honours. He was tragically killed on 23 August 1947 in a crash during the take-off of the prototype Avro Tudor from Woodford airfield. The accident was due to an error during overnight servicing in which the aileron cables were mistakenly crossed.

Chadwick's greatest triumph was, of course, the Avro Lancaster, of which over 7,000 were built. The Lancaster was a truly great aircraft, and is discussed in more detail in Chapter 8. Of the many stories of Lancaster crews continuing their operation on fewer than four engines, one tale gives an appreciation of just how important the Merlin engine was to the aircraft.

Returning from a bombing raid on Hanover on 28 September 1943, Pat Burnett, Lancaster pilot and commander of 9 Squadron, had completed most of the homeward journey across the North Sea on two engines only. The other two had been damaged by enemy fire. As he approached the home airfield at Bardney in Lincolnshire a third engine failed, leaving Burnett with only one Merlin operative to land. Despite having to deal with a crosswind, Burnett successfully landed the Lancaster on his one remaining engine.[47]

That landing must have been an interesting time for the aircrew!

Another great designer was Sidney Camm, designer of the RAF's mainstay in the Battle of Britain, the Hawker Hurricane fighter.

Sidney Camm

Sidney Camm joined the Hawker Aircraft Company as a senior draughtsman in 1923. He became chief designer in 1925. During his employment at Hawker, he was responsible for the creation of fifty-two different types of aircraft, of which a total of 26,000 were manufactured. Some of Camm's designs included the Tomtit, Cygnet, Hornbill, Nimrod, Hart and Fury. He then began designing the aircraft that would become important to the RAF in World War II, including the Hawker Typhoon, Tempest and Hurricane. The design of the Hurricane moved technology from the biplane to the monoplane fighter. As a result, aided by the improved engine technology available from the Merlin, fighters flew faster and higher than before, and, as a consequence, became more deadly.

Camm famously had a one-track mind, utterly committed to the development of his designs. If he thought anyone on the team was not working hard enough there could be quite a telling off given to the guilty party. He was a difficult man to work for, though by common consent a brilliant aeronautical engineer. Many great designers worked with Camm at Hawker. These included Frederick Page, who designed the English Electric Lightning jet fighter, and Stuart Davies, the chief designer of the Avro Vulcan bomber.

Getting the Hurricane fighter into production in sufficient numbers before the outbreak of war was very much helped by the Hawker engineer, Frank Murdoch. Murdoch had visited the MAN diesel engine factory in Augsburg, Germany, in 1936. He came away impressed by the factory's configuration and set-up for large scale production. This visit proved an eye-opener to Murdoch, and many ideas from the MAN factory were incorporated by the Hawker Aircraft Company into the production of Hurricane fighters. More information about the Hurricane fighter is in Chapter 8.

Camm was knighted for services to aviation in 1953, and was elected President of the Royal Aeronautical Society (RAeS), serving from 1954 to 1955. Since June 1971, the RAeS has held the biennial Sir Sydney Camm Lecture in his honour.

Sidney Camm died in 1966 at the age of seventy-two. Just before he died, Camm had been working on the design of an aircraft to travel at Mach 4, having begun his life in aircraft design with the building of a man-carrying glider in 1912, just nine years after the first powered flight.

Along with the Hurricane fighter, another one of the most successful aircraft of World War II was the Mosquito, designed by Ronald Bishop.

Ronald Bishop

Ronald Bishop was born in Kensington, London, in 1903. He joined the de Havilland Company as an apprentice in 1921, becoming Chief Designer in 1936. He continued working there until 1964. The first aircraft for which Bishop was responsible was the DH95 Flamingo – an all metal monoplane that could carry seventeen passengers. It first flew in December 1938.

The outstanding achievement by Bishop's design office was the Mosquito. It was an unarmed high-speed bomber, which achieved 388mph when first tested – Britain's fastest aircraft at the time. Eventually the Mosquito would fly at over 420mph. Built almost entirely from wood, it became known as the 'Wooden Wonder'. The Air Ministry was not agreeable to the radical and untried idea of an unarmed bomber, and did not fund the design. Air Chief Marshal Sir Wilfrid Freeman of the Air Staff was very interested, however, and supported the concept. As a result of the prevalent scepticism at the time, the Mosquito became known as 'Freeman's Folly'. More detail regarding the Mosquito is in Chapter 8.

After the war, Bishop became Design Director on de Havilland's board of directors. During this time Bishop led the design team that produced the DH106 Comet aircraft, the world's first commercial production jet airliner, in July 1949. Bishop held his post on the de Havilland board until February 1964, when he retired. Later that year he was awarded the Gold Medal of the RAeS. Bishop died in 1989.

Why there are no public houses named after Ronald Bishop, Sidney Camm, Sir Stanley Hooker, or any of the other brilliant engineers mentioned here, all of whom helped to save Great Britain and shape the free world, I do not understand. Perhaps it's time for this to be put right.

As the war progressed, the Allies' bombing campaign against Germany to restrict their war materiel output was escalated. This resulted in large losses of aircraft and crew, mainly because there was no fighter escort available for the Lancasters and other bombers which were at the mercy of attack from German fighters. There was no Allied fighter available with the range to fly to Germany, engage the enemy fighters and return – that is, until the North American Aviation P51 Mustang fighter came into service with the Merlin engine fitted. The Mustang was designed by Edgar Schmued.

Edgar Schmued

Edgar Schmued was an American aircraft designer who led the design team at North American Aviation that designed the P-51 Mustang fighter. His adaptation of the new laminar flow wing concept and other innovations made the Mustang's flying qualities superb, while its performance, though good at lower altitudes, required the addition of the Merlin before it became the pre-eminent fighter produced by the United States during World War II.

Born in Germany in 1899, Schmued decided at an early age to become an aviation engineer, and he embarked on a programme of self-study to achieve his aim. He also undertook an engineering apprenticeship in an engine factory. He left Germany in 1925 and emigrated to Brazil, where he found work with a branch of the General Motors Corporation that was concerned with aviation. As a result of his good work for GM, he was sponsored in 1931 to come to the United States and work for the Fokker Aircraft Corporation, owned at that time by General Motors, based in New Jersey. It was here that Schmued became an aircraft designer. In time GM sold its aviation interests and Schmued found himself designing for the

North American Aviation Company.

The British Purchasing Commission, based in New York, was seeking a high-speed fighter to complement its Hurricanes and Spitfires. The Curtis P-40 Tomahawk fighter was the only aircraft that came close to the BPC requirements. Because the Curtis factories were in 1940 at capacity of production, it was decided to approach NAA and get them to make the P-40 under licence in their factories. Accordingly, NAA was approached by the BPC, and the fable of how the Mustang came to be began with NAA's President, James H. Kindelberger (nicknamed 'Dutch'), asking his Chief Designer, Schmued:

"Ed, do we want to build P-40s here?"

Schmued's answer was "Well, Dutch, don't let us build an obsolete airplane, let's build a new one. We can design and build a better one."[48]

Schmued and Kindelberger convinced the British Purchasing Commission that there was no point in purchasing a plane that would soon be out-of-date. Persuaded by their argument, the BPC gave NAA permission to design a new fighter. Schmued and his design team, often working through the night, produced working drawings for the Mustang as quickly as possible, and the first flight of a P51 Mustang took place in September 1942. The aerodynamics of the Mustang were excellent, due mainly to Schmued's use of the laminar flow design principle.

Laminar flow has been defined as the smooth uninterrupted flow of air over the contour of the wings and other parts of an aircraft in flight. If the smooth flow of air is interrupted over a wing, section turbulence is created which results in a loss of lift and a high degree of drag. An aerofoil designed for minimum drag and uninterrupted flow of the boundary layer is called a laminar aerofoil. Schmued made sure the Mustang was designed to maximise laminar flow, making it, according to the Truman Senate War Investigating Committee reporting in 1944, "The most aerodynamically perfect pursuit plane in existence".[49]

However, the Allison engine in the Mustang was not powerful enough to give the aircraft performance sufficient to overcome enemy fighters. It was the addition of the Merlin engine that gave the Mustang sufficient power to enable its streamlined shape to be used to full advantage. (See R. Harker in Chapter 9.)

Following retirement in 1957, Schmued continued his aircraft design work as an independent authority. He worked as a consultant for the US Department of Defence, for private companies and for the film industry in making aviation related films. He continued working until shortly before his death on 1 June 1985.

L to R: Arthur Rubbra, Gordon Lewis, Sir Stanley Hooker, Frank Wooton (the artist) and Jeffery Quill c.1980

Edgar Schmued

Ernest Hives

Miss Shilling at Brooklands Racetrack c.1935

R. Mitchell and H. Royce c.1929

Ronald Bishop

Stanley Hooker

Cyril Lovesey

Sydney Camm

Wilfrid Freeman

Charles Rolls

Roy Chadwick and Guy Gibson c.1943

Chapter 8

Aircraft

The Merlin powered more than forty different aircraft during World War II. Such was the reliability and adaptability of the engine that it was used in many different types of aircraft, including fighters, bombers, reconnaissance and anti-submarine. This chapter concentrates on those aircraft that were of most importance to the Allies during World War II, and includes a short history of their development and application, and the significance of the Merlin in each aircraft.

A full list of aircraft types that were powered by the Merlin is given in Addendum A at the back of this book.

Spitfire and Hurricane

What can anyone add to the many words said and written about these two great aircraft? They are grouped together here because they are seen as the two aircraft that saved Great Britain in the hour of her greatest need. Together they are 'Land of Hope and Glory', 'Rule Britannia' and 'Jerusalem' all rolled into one. Is there any other country on Earth that has two such evocative and emotive emblems of triumph? The partnership of the Hurricane and Spitfire fighters formed the bulwark of the defence of Britain during the summer of 1940. Both aircraft had entered service with the RAF at similar times, and both carried .303 Browning machine guns and, in later models, .79 inch cannon. Both aircraft were powered by the Merlin.

An example of the power of these two aircraft to elicit admiration and pride in one's heritage occurred recently on a visit to the National Memorial Arboretum in Staffordshire with my wife, an ex-WRNS (Women's Royal Naval Service) Officer. We were enjoying the atmosphere of this special place, and happened to notice a party of RAF personnel about to set off for a corner of the Arboretum to dedicate an RAF memorial. We continued on our separate way, fully expecting at some time to meet up with them as we explored the Arboretum and its thought provoking series of memorials. We had not reached the new RAF memorial before an unmistakeable noise reached our ears: the sound of not one, but two Merlin engines as a Hurricane and Spitfire flew over the RAF party in tight formation at precisely one o'clock as the dedication ceremony came to its climax. Everyone around us, everyone in the Arboretum, stopped and listened; stopped and looked at this magnificent sight. Goosebumps. It was, and always is, an overwhelmingly moving sight.

The Hurricane was a very tough, rugged and workmanlike aircraft, able to take a great deal of punishment and remain flying. It outnumbered the Spitfire by around two to one during the Battle of Britain. Some 14,583 Hurricanes had been built by the end of 1944. By this time it had been reduced to a subsidiary role when compared with the Spitfire, which had overtaken it in numbers employed by RAF squadrons.

Today, a total of thirteen Hurricanes are maintained in an airworthy condition around the world. Many more are maintained in static display or are undergoing restoration.

The Mark I Spitfire is one of the most important military aircraft of all time, and entered RAF service in August 1938. Its fragile, almost dainty appearance belied a performance superior to the enemy aircraft it faced, and with its hard-hitting firepower it was a formidable opponent. It was also, for many pilots, their favourite aeroplane to fly. Following the Battle of Britain, the Spitfire became the backbone of RAF Fighter Command, seeing action in all the major theatres of World War II, including Europe, South-East Asia, the Mediterranean and the Pacific. Spitfires continued to be used until well into the 1950s, using the increasingly powerful Merlin and, later, the Griffon engine.

One way that the Spitfire came to epitomise the battle and survival of Britain during the war in many towns and cities throughout the United Kingdom was through a scheme whereby communities could club together and raise money for 'their own' Spitfire. The stated price at the time was £5,000 per airframe. It has been suggested that the actual cost was probably double that price (£570,000 at 2015 prices). Around 3,000 Spitfires were funded in this way, more than any other aircraft type. Each donated aircraft bore a name suggested by its donor, usually marked in yellow characters on the engine cowling. Some of these had rather quirky associations: one was named *Dorothy* because it was purchased with money raised by subscriptions from women who bore that name. Another, *Gingerbread*, was funded by red-headed men and women and flown by flame-haired Australian pilot, Squadron Leader Keith 'Bluey' Truscott, while a third, *The Dogfighter*, was, believe it or not, a gift from the Kennel Club!

The Spitfire was the only British fighter aircraft to remain in production before, during and after World War II. By the time the last Spitfire was rolled out in February 1948, a total of 20,351 aircraft had been built. Today, over fifty Spitfires are maintained in airworthy condition around the world, with a further 180 maintained in static display or undergoing restoration.

Losses of aircraft during the Battle of Britain are difficult to quantify exactly. Different sources give different figures. During the war the numbers of aircraft that were shot down and destroyed were, for propaganda purposes, inflated or deflated, depending which side was reporting. However, most sources consulted agree that the Hurricane fighter accounted for approximately fifty-five per cent of enemy losses during the Battle of Britain, and the Spitfire accounted for approximately forty-two per cent. The remainder of the Luftwaffe aircraft destroyed was due to anti-aircraft fire.

A comment from Leo McKinstry, journalist and author of books on the Spitfire and Hurricane, encapsulates much regarding the importance of the two fighters, and the Merlin engine that gave them advantage over the enemy:

By preventing the Luftwaffe from gaining air supremacy over southern England, the two legendary fighters destroyed the Reich's hopes of mounting an invasion. But these aircraft would never have achieved that success without the Rolls-Royce Merlin engine. Robust and supremely efficient, the Merlin gave the RAF's fighters the power and performance they needed to defend our skies.[50]

Lancaster

The Avro Lancaster, designed by Roy Chadwick, the Chief Designer for the Avro Company, was a four-engine heavy bomber. Over 7,000 Lancasters were built by Avro at their factories at Woodford and Chadderton in Manchester, and Yeadon in Leeds. Lancasters were also built in Canada by the Victory Aircraft Company in Ontario. The final production version, the Mark VII, was made at the Austin Motor Company factory in Longbridge, Birmingham. Such was the need for maximum output during the war that the Yeadon factory employed 17,500 workers at a time when the population of Yeadon itself was only 10,000.

Avro's test pilot, H.A. Thorn, took the controls for the first flight at Manchester's Ringway aerodrome on Thursday 9 January 1941. The aircraft proved to be a great improvement upon its predecessor, the Avro Manchester, being one of the few warplanes in history to be deemed by its pilots just right from the very start. The only modification introduced after its maiden flight was a change to its initial three-finned tail layout, a result of the design being adapted from the Avro Manchester. On the second prototype and all subsequent production aircraft, it was made to the familiar twin-finned specification.

With its vast capacity, capable of carrying a 22,000lb bomb load, the Lancaster needed a special engine, and, as the epitome of reliability, the Merlin was ideally suited to the task. It was the Merlin that powered the aircraft of the Dam Busters Raid in May 1943, believed by many to be the greatest single RAF exploit of the war and one that symbolised Britain's fight back against Germany.

The Avro Lancaster carried the heaviest bomb loads of the war, including the Barnes Wallis designed ten-ton grand slam earthquake bomb. This bomb was used to destroy German U-Boat pens and other well-defended strategic targets. The typical thickness of concrete guarding a U-boat pen would be in excess of twelve feet, and consequently required something very special to penetrate and destroy the pen. The Lancaster first saw active service in 1942 with RAF Bomber Command, becoming the main heavy bomber used by the Allies, including the Royal Canadian Air Force and squadrons from other Commonwealth and European countries. It was the most famous and successful of all World War II night bombers. Throughout the war, the Lancaster delivered over 600,000 tons of bombs in over 150,000 sorties, thereby severely hampering the Nazi output of war materiel, and certainly shortening the war considerably. Although the Lancaster was primarily a night bomber, it excelled in other roles, including daylight precision bombing. Today there are only two remaining airworthy Lancasters: one in the Battle of Britain Memorial Flight, the other based in Canada. There are around a dozen other Lancasters either on static display or undergoing restoration world-wide.

In his book *Wings on My Sleeve*, Eric 'Winkle' Brown, the test pilot, when asked what were the 'greats' in his golden age of aviation from the mid-1930s to the 1970s, said:

Here is my top twenty list, based on wide handling experience of aircraft of that era and judged by the sheer joy of knowing one was flying a real crackerjack. No 1 – Avro Lancaster – just to sit in the cockpit was sheer joy. It exuded self-confidence.[51]

Praise indeed!

Eric Brown is commonly recognised as the worlds greatest ever test pilot. As a Royal Navy officer, he flew 487 different types of aircraft, more than anyone else in history. And in November 2014, Brown, then aged ninety-five, was the guest for the three thousandth edition of BBC Radio 4's Desert Island Discs. During the programme he said that he still enjoyed driving and had just bought himself a new sports car.

Mosquito

The de Havilland DH.98 Mosquito was a British twin-engine multi-role combat aircraft. It served during and after World War II. The Mosquito design team was led by Ronald Bishop, who was Chief Designer at the de Havilland Company between 1936 and 1946.

Constructed almost entirely of wood, it was a very versatile aeroplane used as a medium altitude tactical bomber, a high altitude night bomber, day or night fighter, maritime strike aircraft, photo-reconnaissance aircraft, and as a pathfinder for heavy bomber squadrons.

The Mosquito was made from sheets of Ecuadorean balsa wood sandwiched between sheets of Canadian birch, but in areas needing extra strength, stronger woods replaced the balsa wood filler. The overall thickness of the birch and balsa sandwich skin was only 7/16 inch. This sandwich skin was so stiff that no internal reinforcement was necessary from the wing's rear spar to the tail bearing bulkhead. The Mosquito was not all wood, of course. Metal was used in the construction for engine mounts and other great load-bearing areas of the airframe. However, the total weight of metal castings and forgings used in the aircraft was only 280lb.

Production of the Mosquito began in 1941, at which time it was the fastest operational aircraft in the world with a maximum speed of 415mph, a range of 1,500 miles and a service ceiling of 37,000ft. The Mosquito was, from 1943, formed into the Light Night Striking Force of the RAF, which comprised up to nine squadrons of mosquitos that carried out many bombing raids over Germany. By the end of the war, the LNSF had achieved over 27,000 sorties against the enemy. Mosquitos were capable of carrying quite heavy payloads, frequently being used to deliver blockbuster bombs which weighed over 4,000lb.

The Mosquito was very successful as a bomber, and was the fastest wartime aircraft for over two years in the latter stages of World War II. The concept of a fast unarmed bomber was amply justified in practice with low loss rates as a result of enemy action. Perhaps one of the greatest compliments paid to the Mosquito was from Reichsmarschal Herman Goering when he said:

It makes me furious when I see the Mosquito. I turn green and yellow with envy. The British, who can afford aluminium better than we can, knock

> *together a beautiful wooden aircraft that every piano factory over there is building, and they give it a speed which they have now increased yet again. What do you make of that? There is nothing the British do not have. They have the geniuses, and we have the nincompoops.*[52]

Over 7,781 Mosquitos were built, mostly in the United Kingdom. In the de Havilland factories in Toronto, Canada, however, some 1,133 Mosquitos were assembled and subsequently delivered to the United Kingdom, while in Sydney, Australia 212 were completed by the end of the war. The Mosquito was used by the RAF, Royal Canadian Air Force, Royal Australian Air Force and the United States Army Air Force.

P-51 Mustang

Once Britain was safe from invasion, attention turned to striking back at the enemy through Bomber Command, and from late 1941 the Lancaster gave the RAF the ability to hit the Third Reich hard. The RAF Lancaster squadrons concentrated mainly on night attacks where the darkness helped to protect the aircraft. However, night attacks made accurate bombing very difficult.

The US Air Force decided to attack Germany in daylight in order to improve the accuracy of their bombing raids, but losses of aircraft became unacceptably high. The simple answer to this problem was that the daylight bombing raids needed fighter escorts for protection from German aircraft: fighters that could fly to Berlin and return; long range fighter aircraft that could escort

Allied bombers all the way to Germany and take on the Nazi fighters when they got there. Such an aircraft was developed by the Allies and it was powered by the Merlin.

The Mustang long-range fighter was a single-seat fighter and bomber used during World War II and in the Korean conflict. The Mustang was designed and built by the North American Aviation Company (NAA) in the United States in response to a specification issued to the NAA by the British Purchasing Commission – an organisation, based in New York, which arranged for the purchase and production of armaments from North American manufacturers. The BPC paid for these purchases using Britain's Gold Reserves.

In April 1942, the RAF's Air Fighting Development Unit (AFDU) tested the Allison V-1710 powered Mustang fighter and found it impressive at low altitudes, but not as impressive at higher altitudes. Wing Commander Ian Campbell-Orde, the Commanding officer of the AFDU, invited the Rolls-Royce chief test pilot, Ronald Harker, to try it out. He took the Mustang up for a thirty-minute flight, and after giving it some thought, he sat down with pen and paper on 1 May 1942 and wrote the words that would alter the Mustang's history:

This aircraft could prove itself a formidable low- and medium-altitude fighter. The point which strikes me is that with a powerful and a good engine, like the Merlin 61, its performance could be outstanding, as it is 35mph faster than the Spitfire V at roughly the same power.[53]

It was evident to Harker that the Mustang's performance, although outstanding up to 15,000ft, was inadequate at higher altitudes. This deficit was due largely to the single-stage supercharged Allison engine, which lacked power at higher altitudes. Nonetheless, the Mustang's advanced aerodynamics showed to advantage, as it was about 30mph faster than contemporary Curtiss P-40 fighters using the same Allison power plant. The Mustang Mark I was 30mph faster than the Spitfire at 5,000ft and 35mph faster at 15,000ft, despite the Spitfire having a more powerful engine. The Mustang needed the latest Merlin, the Mark 60 series – two-stage two-speed supercharged versions of the engine – to realise its full potential. Following Harker's comments, Rolls-Royce engineers rapidly concluded that the Mustang powered by a two-stage Merlin 61 would indeed result in a significant improvement in performance at altitude, and started converting five Mustangs to Merlin power as the 'Mustang Mark 10'. This was achieved with a minimum of modification to the engine bay, the Merlin engine fitting well into the adapted engine compartment.

A smooth engine cowling with a 'chin-type' radiator (a small radiator mounted underneath the engine) was tried out in various configurations. What was produced in return was a stunning improvement in performance. Top speed rose to nearly 440mph and climb rates improved dramatically. The Mustang had been transformed from a medium-altitude fighter-bomber to a fully-fledged escort fighter. Enemy aircraft, such as Germany's Me109 and FW190, were equalled and bettered.

Progress was made in getting the Air Ministry on side

when Rolls-Royce's Ray Dorey, head of the Rolls-Royce engine flight test section at Hucknall, steered Harker towards Air Chief Marshal Sir Wilfrid Freeman, who had recently re-joined the Ministry of Aircraft Production (MAP), and who agreed with Harker's view. More than any other person, Sir Wilfrid Freeman (see Chapter 4), in his role as Chief of the Ministry of Aircraft Production during most of World War II, was responsible for the RAF ordering the Hurricane, Spitfire, Mosquito, Lancaster, Halifax and Tempest. It was his role in the development of the Merlin-powered Mustang that was vital. His department, through the British Direct Purchase Commission in New York, provided North American Aviation with the specification for the Mustang.

From late 1943, Mustangs were used by the USAAF's Eighth Air Force to escort bomber raids over Germany. Meanwhile, the USAAF's Ninth Air Force and the RAF used Merlin powered Mustangs as fighter-bombers. In these roles the Mustang contributed greatly to ensuring allied air superiority over the Luftwaffe in 1944. In both the European and Pacific theatres of war, the Mustang claimed 4,950 enemy aircraft destroyed.

Today, Mustangs are coveted by pilots and racers all over the world. Around 200 are maintained in flying condition, mostly in the United States. On the rare occasions when a Mustang comes up for sale, it usually attract bids in excess of $1 million. Over 15,000 Mustang fighters were built before production ceased in 1946.

Chapter 9

Pilots

There are many histories and biographies that tell the stories of the World War II operations carried out by the pilots of the Royal Air Force. The following offers a brief background to some of those pilots in order to give the reader an insight into the extraordinary bravery and indomitable spirit of those involved. The Merlin engine is the common thread that runs through all these biographies.

George Stainforth

Wing Commander George Stainforth was born in 1899. He joined the RAF in March 1923, and in 1928 he was promoted to Flight Lieutenant, receiving a posting to the Marine Aircraft Experimental Establishment, a British military research and test organisation near Felixstowe. It was here that Stainforth became involved in the RAF High Speed Flight.

Following the Schneider Trophy triumph on 16 September 1931, Stainforth had the chance to break the world airspeed record. His first attempt was made in S.6B number S1596, in which he achieved 379mph.

Unfortunately, following a taxiing accident during testing, the S1596 turned over and sank. Stainforth escaped injury. Undaunted, he transferred to S.6B number S1595, which was also fitted with the same specially prepared Rolls-Royce R engine, for the record attempt.

Starting the R engine was not easy. There was considerable risk of an engine explosion because of the special fuel mixture of petrol, methanol and ethyl used. Fortunately no explosion occurred, and on 29 September 1931, Stainforth took off from the water. After a perfect record run over the four timed miles in opposing directions, the record was established at a height of 1,300ft. Stainforth's average speed was 407.5mph. For this achievement, he was awarded the Air Force Cross in October 1931.

Stainforth later went on to break another world record, in typically British eccentric style, by flying upside-down for twelve minutes in the S.6B. Today, incidentally, that record is held by the American pilot Joann Osterud, who in 1991 flew upside-down for four hours, thirty-eight minutes and ten seconds. The things people do to get their name in the record books!

In January 1940, Stainforth was promoted to Wing Commander, commanding RAF 600 Squadron. In June 1940, both Stainforth and Wing Commander Roland Stanford-Tuck were posted to Farnborough, and were given the task of comparison testing the Spitfire with a captured German fighter, the Messerschmitt 109E. Stanford-Tuck was a Battle of Britain fighter pilot and test pilot who became one of the most highly decorated RAF pilots of World War II, credited with shooting down thirty enemy aircraft. Tuck flew the Spitfire, Stainforth flew the

Me109E. Their antics in the two aircraft were, according to some accounts, quite exciting to watch.

After being posted to the Middle East as Officer Commanding RAF 89 Squadron, Stainforth was killed in action on 27 September 1942 while piloting a Beaufighter near the Gulf of Suez. He is buried in the British Military Cemetery at Ismailia, Egypt.

Another member of the RAF High Speed Flight that secured the Schneider Trophy for Great Britain was the magnificently named Augustus Henry Orlebar.

Augustus Orlebar

Augustus Orlebar was born in 1897, and served in both World Wars. He was awarded the AFC in 1921 and Bar in 1929. Between the wars, he was given the command of the RAF High Speed Flight that secured the Schneider Trophy for Great Britain. He became RAF Director of Flying Training in 1940, and subsequently in 1941 he became Air Officer commanding No 10 Fighter Group. In March 1943 he became Deputy Chief of Combined Operations. He died of natural causes in hospital in August 1943 at the early age of forty-six.

It was as a result of the courage and skill of pilots like Orlebar and Stainforth that the performance of the Rolls-Royce R engine was fully realised. The development of the Merlin in time for World War II and the Battle of Britain could not have taken place without their success and the knowledge gained from racing with the R.

Squadron Leader Orlebar, who had shared with R.J. Mitchell the triumphs and setbacks of the previous years, later wrote in his book *Schneider Trophy*:

The credit belongs to the brains which conceive, not to the hands which hold. But the hands had very good fun.[54]

A typically British self-effacing remark.

Guy Gibson

Gibson was one of the most decorated airmen in the RAF during the war (VC, DSO & Bar, DFC & Bar), and is famously remembered as the leader of the Dam Busters raid by RAF 617 Squadron in 1943. He was born in 1918 in Simla, India, the son of a British civil servant. Following the separation of Gibson's mother and father when he was six years old, Gibson and his mother returned to England where he was educated at Earl's Avenue School in Folkestone, Kent, and St Edward's School, Oxford.

Gibson applied to the RAF, but was rejected when he failed the Medical Board. The probable reason for this was that his legs were considered too short. A later application was successful, however, and his personal file includes the intriguing remark: 'Satisfactory leg length test carried out'. He commenced a short service commission in November 1936. After training, he was posted to RAF 83 Squadron. Gibson was in action with 83 Squadron on the day war was declared, 3 September 1939, when he was sent out to locate and attack the German fleet in the North Sea. The attacking force of bombers (Hadley Page Hampdens) found nothing and returned home after jettisoning their bombs in the sea.

Because of the long 'Phoney War' between Germany and the Allies, which lasted from September 1939 until

May 1940, during which there was a lack of major military operations between opposing forces, Gibson did not fly his next operational sortie until April 1940. After completing his first tour of operation, and receiving the Distinguished Flying Cross in the process, Gibson took on a post as a flying instructor. This did not last long, mainly because the lure of operational flying was too great for him to resist. Accordingly Gibson was moved to RAF 29 Squadron at RAF Digby, where he flew Bristol Beaufighters on night-time raids. After flying ninety-nine operations for 29 Squadron and completing his second tour of operation, Gibson was promoted to Squadron Leader and received a bar to his DFC. Again, he returned to flying instruction, but was soon back in operations at RAF Coningsby in 1942. He arrived there just as 106 squadron was having its Manchester bombers replaced by Avro Lancasters. Gibson completed a third tour of duty with 106 Squadron and was promoted to Wing Commander, adding a Distinguished Service Order and Bar to his medals tally when he was only twenty-four years old.

Because of Gibson's formidable record, Marshall of the RAF, Sir Arthur Harris, chose him as the first commanding officer of RAF 617 Squadron, specially formed to carry out the bombing of the Ruhr Dams in Germany using the bouncing bomb developed by Barnes-Wallace. Gibson was allowed to choose the pilots and aircrew he wanted in the Squadron. After the Dam Busters raid, Gibson received the VC; he was now a national hero and the most highly decorated pilot in the RAF.

Air Vice Marshall Cochrane and Harris now wanted Gibson to remain on the ground. It would have been a blow to morale if he had been shot down or captured by the Nazis, so he was given a non-operational post at a bomber base in Lincolnshire. This, to a man like Gibson, was not enough. He pestered both Cochrane and Harris to let him return to operations. They reluctantly agreed and gave him a bomber squadron of Mosquitos to lead into Germany.

On 19 September 1944, Gibson led his squadron to attack industrial targets in Rheydt and Monchengladbach. He did not return. His Mosquito crashed at Steenbergen in Holland. The cause of the crash has never been satisfactorily explained.

Harris and Cochrane regretted their decision. Britain had lost one of its greatest heroes. Harris described him as: "As great a warrior as this island ever produced".[55]

Guy Gibson was twenty-six years, one month and seven days old when he died.

Douglas Bader

Douglas Bader was born in 1910 in St Johns Wood, London. He was the second son of Frederick Roberts Bader, a civil engineer, and his wife Jessie. Bader's father saw action in World War I with the Royal Engineers, and was wounded in action in 1917. He stayed in France after the war, where he died of complications from his war wounds in a hospital in Saint-Omer in 1922.

Bader joined the RAF in 1928, and was commissioned as a pilot officer into No 23 Squadron, based at Kenley in Surrey, on 26 July 1930. Bader became a daredevil while training there, often flying illegal and dangerous stunts. A strict order was issued forbidding unauthorised aerobatics below 2,000 feet; Bader took this as an unnecessary health and safety rule rather than an order to be obeyed. On 14 December 1931, while visiting Reading Aero Club, he attempted some low-flying aerobatics at Woodley Airfield in a Bristol Bulldog Mark IIA of 23 Squadron, apparently as a dare. His aircraft crashed when the tip of the left wing touched the ground. Bader was rushed to the Royal Berkshire Hospital where both his legs were amputated – one above and one below the knee.

In May 1933 Bader was invalided out of the RAF against his will, in spite of the fact that, after recovering from his crash, he had retaken flight training and passed his flight checks. Bader took an office job with the Asiatic Petroleum Company, later to become known as the Shell Oil Company.

At the outbreak of war in 1939, Bader returned to the RAF as a pilot and took part in the Battle of Britain. Bader was posted to command No 242 Squadron, a Hawker Hurricane unit based at RAF Coltishall, as acting Squadron Leader on 28 June 1940. No 242 Squadron comprised mainly of Canadian pilots who had suffered high losses in the Battle of France and were experiencing low morale. Despite initial resistance to their new commanding officer, the Canadian pilots were won over by Bader's strong personality and perseverance, especially in cutting through red tape to make the squadron operational again. No 242 Squadron became fully operational on 9 July 1940, Bader having transformed it into an effective fighting unit, successful in the defence of Britain during the Battle of Britain.

Bader was shot down over France on 9 August 1941 while flying a Spitfire on an offensive patrol over the French coast. His Spitfire was badly damaged and he lost height rapidly. He jettisoned the cockpit canopy and released his harness pin, but his prosthetic leg was trapped. Part way out of the cockpit, and still attached to his aircraft, Bader released his parachute, at which point the retaining strap on his leg snapped under the strain and he was pulled free. He was subsequently captured and sent to a prisoner of war camp.

Having now lost one of his legs, Bader was supplied with a replacement leg when it was arranged between the German forces and the RAF that it be dropped by the RAF over Saint-Omer Airbase in France. Despite his disability, he made a number of escape attempts, and after frustrating the Germans, he was sent to the supposedly 'escape-proof' prison at Colditz, where he remained until April 1945 when the prison was liberated.

For Bader to lose both legs and still achieve so much is awe-inspiring. It has been said that Bader's lack of legs contributed to his success as a fighter pilot. When in dogfights with the enemy, pilots were subjected to high G-forces that often caused them to 'black out' and lose consciousness as the flow of blood from the brain drained to other parts of the body, usually the legs. As Bader did not suffer this disadvantage, he could remain conscious longer, and thus had a benefit over more able-bodied opponents.

Incredibly, Bader was able to continue playing golf well into later life, and achieved a handicap of four. A golfing companion noticed how Bader always seemed to hit the ball better off an uphill lie. Accordingly, Bader had a spare prosthetic leg made slightly shorter than the other and wore it when playing, giving him the advantage of always playing as if the ball was on an uphill lie.

After the war, Bader joined the Shell Oil Company where he continued flying, becoming Managing Director of Shell Aircraft until he retired in 1969. Douglas Bader died on 5 September 1982 in Chiswick, London, aged seventy-two. He has a public house named after him in Ipswich, Suffolk.

Ronald Harker

Ronald Harker was born in 1909. When war started in 1939, he was already a seasoned pilot, and he became the senior test pilot for Rolls-Royce. During the Battle of Britain, Harker flew from combat squadron to combat squadron, organising the flow of spare parts to keep the RAF fighters flying. At the same time he stood ready to defend the Rolls-Royce experimental aerodrome at Hucknall near Nottingham from German bombers. Harker test-piloted the Merlin 60 series of engines at Hucknall. These engines eventually powered the Spitfire Mark IX, crucially helping to maintain the technical lead over Germany after the Battle of Britain.

Harker was the son of the Chief Medical Officer for Tyneside. He joined Rolls-Royce as an apprentice, but when the worldwide economic depression of the 1930s hit Rolls-Royce, he was laid off work, though he continued to fly as a club pilot. Harker re-joined Rolls-Royce in 1934 and became the company's Chief Test Pilot at Hucknall Aerodrome, where he tested and flew aircraft powered by Goshawk, Kestrel and Merlin engines.

Harker's place in the history of aviation is assured.

He is best remembered as the man who put the Merlin in the Mustang, as he was the first to spot the potential of a Merlin-powered Mustang when he tested an early production aircraft in April 1942 at RAF Duxford, and commented on the underpowered Allison engine (see item on the Mustang). Harker knew that the Mustang would be a world-beater with the latest Merlin 61 engine, and he convinced Rolls-Royce and the Air Ministry to try it out. Six weeks later, in June 1942 the first Merlin-powered Mustang was flown from Hucknall. It was an immediate success, and went on to become the most successful fighter of the latter stages of World War II. Its range was such that it accompanied US Eighth Air Force heavy bombers on operations over Berlin, where it would engage with enemy fighters and defend the bombers. Harker was given a pay rise of one pound a week for his efforts (equivalent to £33 in 2015). He left Rolls-Royce in 1971, and retired to New Zealand, where he died in 1999.

The Memorial Window

In January 1949 a Battle of Britain commemorative window was unveiled by the Chief of Air Staff, Marshall of the Royal Air Force, Lord Tedder GCB. Its location was the Marble Hall entrance foyer to the Rolls-Royce Main Works in Nightingale Road, Derby, and it was an impressive and emotive sight. Following the closure of the Derby Main Works in 2007, the window was put into storage. After restoration, it now forms a centrepiece at the Rolls-Royce Learning and Career Development Centre at Wilmore Road in Derby.

The window contains the following quote:

> *This window commemorates the pilots of the royal air force who in the battle of britain turned the work of our hands into the salvation of our country.*[56]

The commemorative window is the work of Hugh Easton, who also designed the Battle of Britain window in Westminster Abbey. Easton described the Derby window in these words:

In the centre of the window stands the figure of a typical fighter pilot of the Royal Air Force. Ready for battle, in flying boots and 'Mae West', his helmet in his hand, he stands on the spinner of an airscrew, its three blades dominating the lower part of the window. Behind it are stretched out in long lines the sheds and buildings of the Derby factory which produced these engines with which the pilots won the Battle of Britain. In the lower part, therefore, I have tried to symbolise the work of man's hands; the machine or, one might say, the structure: the bones and muscles. In the centre, I felt, the pilot should represent the brain. Above and behind him, with outstretched wings, ready to strike, is a golden eagle; here is the heart and spirit. Beyond, and framing the eagle and dominating all the top of the window, is the resplendent sun in all its glory, symbol of that for which the Battle was fought, and towards which humanity lifts up her eyes.[57]

The window also commemorates another battle – the battle for output, for production. While those RAF pilots fought over England, the workforce of the Derby factory increased between January 1940 and September 1940 from 13,000 to 17,500. The numbers of Merlin engines produced increased from 350 to 750 per month. And, as we have seen, other factories exceeded the output of the Derby factory by a considerable amount. The exhortation of Rolls-Royce General Manager, E.W. Hives, to "Work Until it Hurts" was obeyed by the Rolls-Royce workforce as if it were an order given on the field of battle. Hives, modest as ever, said at the opening ceremony of the commemorative window:

> *We workers at Rolls-Royce like to consider ourselves part of the RAF. We recognise that our efforts were more congenial and we were not called upon to display the courage and suffer the hardships and sacrifices of the fighter pilots. Nevertheless, we hope we shall not be considered impertinent if we insist on our claims that we were a part of the RAF in that battle. We would like it to be thought that where the inscription on the window reads 'our hands' this covers not only the workers in the Rolls-Royce factories, but all those on the industrial side who made their contribution.*[58]

When one considers the extremely long working hours put in by the Merlin production workforce in all of the main factories, and by their legion of suppliers, and the constant risk of aerial attack on those factories and workplaces from the Luftwaffe, those workers were certainly not just part of the RAF, but an indispensable part of the effort to defeat Nazism. It could be argued that victory on the production line was at least as important as victory gained in all other areas.

Chapter 10
More Than an Aero Engine and Beyond

The Merlin was not only used as an engine to power aircraft. It was modified and developed in order to be used in other wartime roles, including being used in British Army tanks and Royal Navy marine craft. It was the advent of Sir Frank Whittle's gas turbine engine that signalled the beginning of the end of the internal combustion engine as a power unit for military aircraft.

Meteor Tank Engine

In 1940, the British Leyland Company was manufacturing a tank under orders from the Government's Mechanisation Board. The company was dissatisfied with many features of the design of the tank, and had little confidence in the Meadows flat-12 engine which they had been ordered to manufacture for the tank. Henry Spurrier, a top executive at British Leyland, approached W.A. Robotham at Rolls-Royce, who at that time was in charge of the Chassis Design and Development Division based in Belper near Derby, hoping for some technical assistance. Following Spurrier's approach, Robotham took the concerns to his boss, Ernest Hives. The outcome of that meeting was to combine the engineering expertise of Rolls-Royce and British Leyland to see if the many problems which British tanks were experiencing could be resolved. Robotham and his team were given the task of investigating and understanding the problems. Meanwhile, the engineers at British Leyland were busy converting their factories from truck to tank production. Robotham's team remit was to produce a new, more powerful and reliable power plant and to improve the steering, transmission and suspension of a suitable existing model of tank. It was decided to focus improvements on the Mark VI Crusader tank.

The engine that was to power this improved tank had to run on 'pool petrol', that is ordinary petrol, and not the high-octane fuel which was the sole preserve of the RAF. The Merlin aero engine was chosen as the best engine to go into the tank. One of the main reasons for this was that Rolls-Royce and its supply chain were well geared-up to produce Merlin parts in exceptional quantities. In order to transform the Merlin into a reliable tank engine, the supercharger was replaced by conventional carburettors. Other modifications included reversal of the engine's direction of rotation to comply with automotive practice, and the greater use of cast-iron and steel component parts, which were easier and cheaper to manufacture than the exotic lightweight alloys used in the Merlin. The reduction gears and other related equipment were also removed. This greatly simplified the construction of the new tank engine that became known as the Meteor. It was, in effect, an un-supercharged Merlin that was also de-rated, or detuned, causing a reduction in its power output down to 600hp, which was more than enough to power a tank. The engine did operate using pool petrol and proved exceptionally reliable. For the first time, British tanks had plenty of dependable power.

In common with millions of other workers in industry in Britain throughout the war, the design and development team on the Meteor engine worked twelve hours a day throughout the week, and eight on Sundays "when they had a break at noon to drink draught Bass at the White Hart, Duffield".[59] The White Hart is still there today, though in common with many other hostelries in Britain it is now a pub-restaurant.

The first Meteor engine was tried in a modified Crusader tank in September 1941 at Aldershot. However, despite the design and development work undertaken by the Rolls-Royce team based at the Clan foundry near Belper, and despite the efforts of the British Leyland workers in getting the prototype engines installed, a major setback occurred just six months into the project. Leyland let it be known that they wished to back out of the partnership. They would be willing to act as sub-contractors, but they did not wish to take responsibility for the engine.

Robotham took the news from Leyland to Hives, and after a series of meetings with the Ministry of Supply, it was decided to cancel the order for tank engines from Leyland and to give Rolls-Royce the job of supplying the engines. Hives then went to see Lord Beaverbrook at the Ministry of Supply. He told Beaverbrook that he already had his hands full making Merlins for the RAF, and if Rolls-Royce were to make tank engines they would want £1 million up front to finance the project and provide the necessary tooling, equipment and premises. They would also accept no interference from Government during the development of the tank engine. After their meeting, Beaverbrook telegrammed Hives with the following message on 12 September 1941:

OHMS Ministry of Supply

E.W. Hives Nightingale Road Rolls-Royce Derby

The British Government has given you an open credit of one million pounds stop
This is a certificate of character and reputation without precedent or equal stop

Beaverbrook.[60]

Because of its massive programme of production of the Merlin aero engine, Rolls-Royce agreed to be the main source of supply of engine parts for whomever replaced the Leyland Company as main manufacturer. Eventually, the Meteor was mass produced by the Henry Meadows Company Ltd in Wolverhampton, and subsequently by the Rover Car Company, who took over production of the Meteor in 1944. The first extensive use of the Meteor was in the Cromwell tank, which eventually replaced the Crusader tank in 1944 and was used during D-Day and in Northern Europe. Later the Government made Rover responsible for the research and development of non-aero large military engines. Rover continued in this role long after the war, producing various derivatives of the Meteor.

Production of the Meteor by Rover ceased in 1964, by which time over 9,000 engines had been produced for the British armed forces.

From these beginnings, the Meteor tank engine was developed, manufactured and operated in various tanks throughout the war and post-war. The poor reliability of British tanks became a thing of the past. The driving force behind many of these improvements, W.A. Robotham, was made Chief Engineer of Tank Design and joined the Government's Tank Board.

Marine Merlin

The Merlin was used by high-speed ocean-going craft, which included motor torpedo boats and motor gun boats for the coastal forces of the Royal Navy. The marine version was also used in RAF air-sea rescue launches. For all these vessels, the Merlin was re-engineered for use in a marine environment, including the use of a single stage supercharger. Seventy such craft were delivered to the Royal Navy before the project was abandoned to give priority to the production of aero engines.

Beginnings of Gas Turbine Power

The arrival of Frank Whittle's gas turbine engine signalled the beginning of the end for the internal combustion engine in the fighter aircraft of the world's air forces. Subsequently, the new jet-engine technology took over from engines like the Merlin and Griffon, and continues to this day. How the Rolls-Royce Company became involved in gas turbine development and manufacture is quite fascinating.

In 1942, Stanley Hooker knew that Frank Whittle was

developing a gas-turbine engine, and he wanted Rolls-Royce to be involved. Hooker managed to convince his boss, Ernest Hives, that a visit to Whittle's workshops, in a disused foundry in Barnoldswick, Lancashire, was worth a try. Whittle was glad of the interest shown, and took the two on a tour of his workshop, during which Hives asked where all the engines were.

The response from Whittle was "Well, we've only got two – can't get the parts made."

Hives responded, "Send us the drawings. We'll make the bits for you."[61]

The Ministry of Aircraft Production had chosen the Rover Car Company to make Whittle's jet engines. The relationship between Whittle and Rover was not working, however, and communications between the two were strained. Hooker had heard about this conflict between Whittle and Rover, and he was thinking of a way to secure the engine's development for Rolls-Royce. He wanted to make that jet engine. A meeting was arranged, and Hooker and Hives met S.P. Wilkes, the Chairman of the Rover Car Company, for dinner at the Swan and Royal Public House in Clitheroe, Lancashire. At the conclusion of the meal, Hives said to Wilkes:

"Now what are you doing with this jet? You shouldn't be doing that engine. I'll tell you what – I'll give you our tank engine factory in Nottingham, and you give me this jet job."

Wilkes looked up and said, "Done."[62]

A handshake closed the deal that ultimately helped transform Rolls-Royce into the global company it is today. During this conversation, there was no talk of money, and no talk of getting Government agreement to the arrangement. Hives and Hooker had obtained an important addition to the armoury of Rolls-Royce, but what made it possible was the astonishing sense of cooperation and reliance that existed in British industry during the war. Companies that had been fierce competitors prior to the war found themselves working together and sharing their secrets. The exchange deal between Rover and Rolls-Royce became official on 1 April 1943. Subsequent Rolls-Royce jet engines would be designated in an 'RB' series, RB standing for Rolls Barnoldswick, an example being the Derwent jet engine becoming the RB26.

Rolls-Royce thereby acquired jet engine development and the Barnoldswick factory where the engines were being made by Whittle, and became responsible for the production of jet engines. That the three, Hives, Hooker and Wilkes, met for dinner in Clitheroe is not in doubt. However, the sealing of the agreement in such a short time tends to be questioned sometimes. Memoranda and other records prove that there were a number of notes between the protagonists around that time, discussing the situation between Rover and Whittle and the involvement of Hives, and certainly the characters would have been in contact before the dinner took place.

I am inclined towards Hooker's version of the sealing of the arrangement. He had no reason to exaggerate.

Chapter 11
Today

Merlin Repair and Overhaul

Today in Britain, there are over seventy airworthy Merlin-powered World War II aircraft, including the Spitfire, Hurricane, Lancaster bomber, Mustang fighter and Mosquito. Some are in the Battle of Britain Memorial Flight, others in business organisations that own restored aircraft, and some are in private ownership. Many more aircraft from World War II are under restoration or on static display.

The numbers of Merlin-powered airworthy aircraft throughout the world is in excess of 300, many of which are Mustang fighters based in the United States. The Merlin engines used in these aircraft require overhaul, servicing and restoration to maintain their airworthiness. In Britain at the time of writing two of the most well-known engineering organisations that repair and service Merlin engines are Retro Track and Air Ltd and Eye Tech Engineering Ltd.[63]

A recent visit to the Retro Track and Air Company proved to be an interesting experience. The business has a main entrance frontage of Cotswold Stone, and the entrance reception area has a Merlin engine in gleaming condition on permanent display. Discussion with the factory staff indicated that they are very aware of the heritage that they are working to preserve and maintain.

A tour of the manufacturing shop floor showed a number of aero engines in various stages of repair and overhaul, and the machining of Merlin components, including cylinder heads that had been cast and supplied by a British company. The engines included the Merlin and Griffon engines, and some Bristol Aeroplane Company engines, the Cherub and the Mercury. Work was also just starting on the repair and overhaul of Merlin radiators and oil coolers.

On-site in the factory is a fuel room where testing is carried out on carburettors. Exhaust stubs for a Merlin engine were being made prior to the engine being fitted in a Lancaster bomber. There is also a large component store that carries just about every spare component part imaginable. There appears to be no part of the Merlin engine that Retro Track and Air are not able to repair or replace, including the electrical components such as magnetos. And just to prove that the business is not confined to aero engines, a Spitfire tail plane was being repaired in the workshop prior to being returned to the customer for fitting.

The Retro Track and Air Factory is also equipped to carry out all types of non destructive testing, including Magnetic Particle, Ultra Sonic, Eddy Current and Dye Penetrant Inspection. These vital techniques of inspection are used to ensure that the metal used in engine components is not cracked or defective in any way, imperative to the determination of good quality components. Retro Track and Air are Ministry of Defence and Rolls-Royce approved to carry out work on Merlin, Griffon and other engines.

Battle of Britain Memorial Flight

The Battle of Britain Memorial Flight (BBMF) is based at RAF Coningsby in Lincolnshire. It provides aerial display through the use of Supermarine Spitfire, Hawker Hurricane, Avro Lancaster, Dakota Transport and the Chipmunk Trainer. The BBMF aircraft are regularly seen at air displays all over the British Isles and Europe, and are used during anniversary commemorations of World War II and at British state occasions, which have included the Trooping the Colour and the royal wedding of Prince William and Kate Middleton in 2011. The visitors' centre at RAF Coningsby is well worth a visit, giving the opportunity to see the BBMF's historic aircraft at close quarters and to observe the BBMF technicians working to maintain them in airworthy condition. At the time of writing there are no aero engines on display at the visitors' centre.[64]

P-51 Mustang

In the United States, surplus Merlin engines and Mustang airframes were sold off relatively cheaply at the end of the war. Subsequently many were modified for competition in the air racing events held throughout the United States. The Mustang and the Merlin are still in use during these air races, and work continues on increasing the power of the Merlin. One of the more famous racing events for World War II fighter aircraft is the annual Reno Air Races, where innovations and modifications to the Merlin have led to large increases in power output. One of the modifications introduced was the replacement of the intercooler with Anti-detonate Injection, using fifty per cent distilled water and fifty per cent Methanol, which is almost identical to the Luftwaffe's wartime MW50 system. Many of the fastest aircraft at these events have increased Merlin manifold pressures to 56psi, equating to 3,800hp, allowing the Mustang to attain speeds beyond 500mph.

The first Reno Air Races in 1964 were organised by World War II veteran Bill Stead. They take place each September at the Reno Stead Airport a few miles north of Reno, Nevada, and regularly attract over 200,000 spectators. Racing aircraft in the Unlimited Class, which consists almost entirely of modified World War II fighters, routinely reach speeds well in excess of 400mph. In 2003, Skip Holm, piloting a modified P-51D Mustang with a Merlin engine, reached an all-time speed record of 507.105mph in a six-lap race around the eight-and-a-half mile course. The continual striving for ever faster speed and power during these air racing events is a tribute to the adaptability and seemingly never-ending potential of the Merlin engine.

Douglas Bader – the war is over!

Guy Gibson with VC Ribbon

Lady Huston c.1909

Lord Beaverbrook

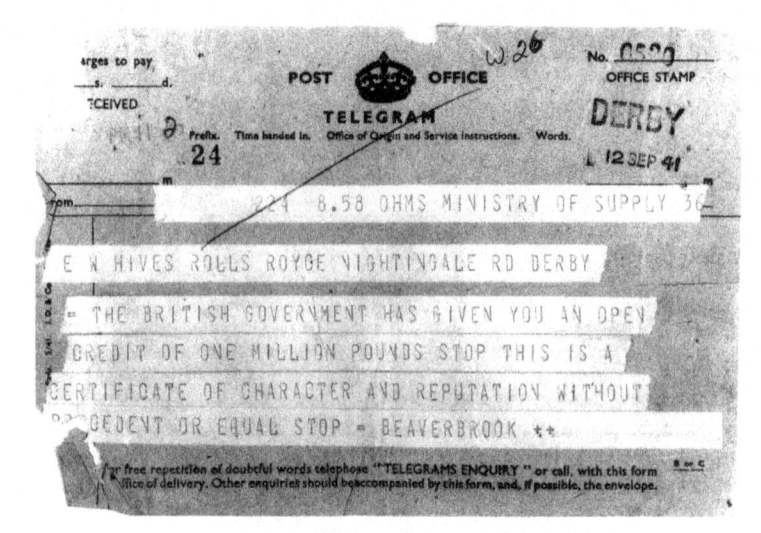

Lord Beaverbrook's telegram to Hives

RAF High Speed Flight 1931
L to R: Flt Lt E.J.L. Hope, Lt RL Jerry Brinton (Fleet Air Arm),
Flt Lt Freddy Long, Flt Lt George Stainforth, Sqn Ldr AH Orlebar
(Flight Commander), Flt Lt John Boothman, Fg Off Leonard Snaith,
Flt Lt W.F. Dry (Engineering Officer)

A Merlin after repair at the Retro Track and Air Ltd factory in Gloucestershire.
April 2014

The Tilly Shilling Public House in Farnborough

Chapter 12
What If?

During the 1930s there had been a number of indicators of an impending conflict. Some people reacted to those indicators by speaking publicly and warning of the imminent war and Britain's lack of preparedness. Winston Churchill was one, but there were many others. One well-known person who was vociferous in his concern for the future security of Britain during the 1930s, and who lost his job as a result, was the editor and writer, Capt. W.E. Johns.

Captain W.E. Johns is famous throughout the world as the author of the *Biggles* series of fictional adventure stories written around the escapades of pilot James Bigglesworth. Johns was a regular contributor to the *Modern Boy* magazine in the late 1930s, and he edited and wrote for the magazines *Popular Flying* and *Flying*. From the early 1930s, Johns was calling for more pilots to be trained in the RAF, because if there were not enough pilots when war came, "training would have to be rushed, and under-trained airmen would be at a disadvantage in combat against better-trained German pilots."[65] Johns wrote acerbic editorials in Popular Flying and Flying magazines which were very strongly opposed to the Government's policy of appeasement to the Nazi regime. Johns was also critical of several Conservative politicians of the time. He attacked the Government's "foul and craven hypocrisy"[66] for its non-intervention policy in Spain during the Spanish Civil War, in which Nazi forces were involved. John's opposition to appeasement is also reflected in some of his books. In Biggles & Co (1936) the storyline revolves around German preparations for war and conquest, and, more advanced in his thinking for that time than most, in Biggles, Air Commodore (1937) he alludes to Japanese preparations for the conquest of British colonies in the Far East.

The Government's reaction to being condemned in the most widely read aviation magazines in the world, magazines that were read by many in the RAF, was to apply pressure on the publishers of the magazines to sack Johns. Johns was indeed sacked as editor of both publications at the beginning of 1939, but he continued writing very successfully. The Air Ministry in 1940 asked Johns to write something that might inspire young women to join the Women's Auxiliary Air Force (WAAF). Accordingly Johns wrote an eleven volume series detailing the exploits of WAAF Flight Sergeant Joan Worralson, nicknamed Worrals. This series of books was published between 1941 and 1950.

The advent of war showed that Johns had been far-sighted and correct to raise his concerns. Germany invaded Poland on 1 September 1939, one week after the signing of the Molotov-Ribbentrop Pact, a non-aggression pact between the Soviet Union and Germany. The Soviets invaded Poland on 17 September, and subsequently the two aggressors divided and annexed the whole of Poland on 6 October.

On 3 September Britain and France, both countries having agreements with Poland to defend her should an aggressor invade, declared war on Germany, realising that Hitler's aims in Europe could only be met and repulsed with force. After the declaration there followed a period that became known as the Phoney War, between September 1939 and May 1940, marked by a lack of major military operations by Great Britain and France against Germany following the invasion of Poland. War had been declared, but no side committed itself to launching a significant land offensive as the protagonists appeared to deliberate about what to do next.

Following this period of relative inactivity and seizing the initiative, Germany invaded Denmark and Norway in April 1940. Denmark succumbed after a few hours, and despite some Allied support, Norway was subjugated in less than two months. The German onslaught continued into May 1940 with the invasion of France, Belgium, the Netherlands and Luxembourg, all of which were quickly overrun. The Nazi forces used *Blitzkrieg*, or 'Lightning War', tactics. By means of speed and surprise, mechanised infantry ground troops, supported by heavy armour and constant close air support, would break through an enemy's line of defence. These tactics proved so successful that the British Army and what remained of the French and Belgian forces were beaten, surrounded and stranded at the French port of Dunkirk by the end of May. This defeat, described by Winston Churchill as a colossal military disaster, led to the 'Miracle of Dunkirk' between 27 May and 4 June

1940. Over 338,000 British and Allied troops were rescued by the Royal Navy and a fleet of over 800 other boats from Britain that included merchant marine craft, lifeboats, pleasure craft and fishing boats. The British Army had to abandon all their tanks, vehicles and other equipment in France and return home a defeated force that was now ill-equipped to wage any land war with Germany.

By the middle of June 1940, with most of Western Europe under Nazi and Italian domination, all that stood between Hitler and the invasion and overthrow of Britain were the Royal Navy and the Royal Air Force. Hitler then decreed that the Royal Air Force had to be wiped out prior to an invasion of the British mainland by German forces.

Hindsight

> *With the benefit of historical hindsight we can all see things which we would wish had been done differently or not at all.* [67]

Some of today's armchair historians might answer the question 'What if Great Britain had lost the Battle of Britain in 1940?' by saying that the Nazis never had command of the air and the sea. They would argue that both would have been necessary for the Nazis to succeed in removing Great Britain from the war, and therefore defeat was very unlikely.

However, if one looks at some of those writing at or near the time of the battle, a somewhat different opinion emerges. Take, for example, the comments of Joseph Walshe, the Secretary of the Department of External

Affairs of the Irish Free State, in a memo to Eamon de Valera, the Irish Taoiseach, in June 1940. This report is not summarised, and is quoted in full. It brings home the thoughts and feelings shared by many people at the time.

The memo is titled *Britain's Inevitable Defeat.*

Britain's defeat has been placed beyond all doubt. France has capitulated. The entire coastline of Europe, from the Arctic to the Pyrenees, is in the hands of the strongest power in the world which can call upon the industry and resources of all Europe and Asia in an unbroken geographical continuity as far as the Pacific Ocean. Neither time nor gold can beat Germany. It is frankly acknowledged in America that America must look to her own defences. She may be at war with Japan in a few short weeks. Senator Pittman, Chairman of the Foreign Affairs Committee of the Senate, expressed the view yesterday to the Press that nothing that America could do could affect anything more than a delay in the final defeat.

England has the most concentrated industry and system of ports of any great power in the world. Her power of production would be wiped out in a few weeks of intensified bombing and her ports put out of action. Italian and German submarines acting in combination are strong enough to throw her merchant fleet into confusion. The German Air Force is acknowledged to have had an immense superiority in numbers, even while France was in the war. Germany is foregoing the use of captured

French planes against England. Britain has suffered a colossal military defeat [at Dunkirk] *and the bulk of her effective forces have been rendered useless for months to come by the loss of the greater part of her war materiel.*

All the smaller states in Europe on which she was relying for incidental support have grown cold and are abandoning her. Rumania is going Axis. Turkey has slipped out of her obligation to take action against Italy. Greece is having friendly talks with the Axis Powers. In Africa, Egypt refused active participation. South Africa is on the verge of Civil War because at least fifty per cent of the population were opposed to participation, and their numbers are now being rapidly added to by the clear evidence of England's approaching defeat. General Hertzog's return to power and South Africa's withdrawal from the war appear to be a certainty. In Asia, Iraq is hesitating about further co-operation and is having consultations with Turkey and Egypt as well as the other Arab States. Japan is setting up an East Asia Monroe Doctrine and has begun an undeclared war against the British in Hong Kong. She is openly threatening an early move against French Indo-China and the Dutch East Indies. No wonder the American radio is sending out rumours of pending peace negotiations through the British Ambassador in Madrid. It is a fair deduction from the course of events that some members at least of the British Cabinet must be turning their thoughts to peace.[68]

When I first read Walshe's comments during research for this book, I was very thankful indeed that in Britain we at least had Churchill on our side.

What might have happened had the Merlin engine not been as successful as it was, if the Battle of Britain had been lost and the Royal Air Force as an effective fighting force had been eliminated by the Nazi Luftwaffe?

Some possible scenarios and their consequences include:

1. Britain suing for peace with Germany and becoming a non-combatant in the war. The British Isles could have remained uninvaded, but it would have been likely that a significant portion of the strategic areas of British influence, such as Malta, Gibraltar and the Suez Canal, would have been given up to the Axis forces.
2. Nazi Invasion of Great Britain and conquest of the British people.
3. Britain resisting invasion and fighting on with a depleted Royal Air Force. Her island and infrastructure would have been at the mercy of German bombing, because the Nazis would then have had superiority in the air, though probably not superiority at sea.

The issue of Great Britain suing for peace in 1940 and becoming a non-combatant in the war was certainly not an impossibility. Some members of the Government of National Unity at the time, notably Lord Halifax and Neville Chamberlain, are recorded as not being against an accommodation with Hitler. The evacuation of Allied troops from Dunkirk had just begun on 27 May, and was to continue until 4 June 1940 before all the British troops were safely home. These were dark days indeed. Churchill called a Cabinet meeting on the evening of 28 May 1940, and began his address to twenty-five senior Government ministers with a summary of the status of the war at that time. Hugh Dalton, the Minister of Economic Warfare, recorded in his diary what Churchill said next:

And then he said, "I have thought carefully in these last few days whether it was part of my duty to consider entering into negotiations with That Man. But it was idle to think that, if we tried to make peace now, we should get better terms than if we fought it out. The Germans would demand our fleet – that would be called 'disarmament' – our naval bases, and much else. We should become a slave state, though a British Government which would be Hitler's puppet would be set up.

"And I am convinced that every man of you would rise up and tear me down from my place if I were for one moment to contemplate parley or surrender. If this long island story of ours is to end at last, let it end only when each one of us lies choking in his own blood upon the ground."[69]

Churchill's words were greeted by loud cries of approval and support from all round the table. There were to be no further thoughts of parley with the Nazis. The will of Winston Churchill and his supporters in the cabinet prevailed, and all of them stood firm against a pact with Germany.

There are, of course, many other possible outcomes, not least of which would be Britain becoming non-combatant followed by the eventual defeat of Germany by the forces of the Union of Soviet Socialist Republics – the USSR. This could well have led to Britain and most of Europe becoming Communist states.

The order 'That Man' gave intending to bring about the end of Great Britain as a fighting force is as follows:

Hitler's Directive Number 16

On 16 July 1940, Hitler ordered the preparation of a plan to invade Britain. It was known as *Directive No 16; On the Preparation of a Landing Operation Against England.* In summary, it read as follows:

Since England, despite its militarily hopeless situation, still has not shown any signs of being prepared to negotiate, I have decided to prepare a landing operation against England and, if necessary, carry it out. The objective of this operation is to eliminate the English home country as a base for the continuation of the war against Germany.[70]

Included in Hitler's directive were the following preparations and commands required to make an invasion possible:

> *The English air force must have been beaten down to such an extent, morally and in fact, that it can no longer muster any power of attack worth mentioning against the German crossing.*[71]

Hitler's directive goes on, and includes the following points:

1 *The landing will be in the form of a surprise crossing on a wide front from about Ramsgate to the area west of the Isle of Wight. Units of the Air Force will act as artillery, and units of the Navy as engineers. Preparations for the entire operation must be completed by the middle of August.*

2 *These preparations must also create such conditions as will make a landing in England possible, viz.:*
  i *Mine-free channels must be cleared.*
  ii *The Straits of Dover must be closely sealed off with minefields on both flanks; also the Western entrance to the Channel approximately on the line Alderney-Portland.*
  iii *Strong forces of coastal artillery must command and protect the forward coastal area.*

3 *Command organisation and preparations. Commander-in-Chief Army will detail one Army group to carry out the invasion. The invasion will bear the cover name 'Seelöwe' [Sea lion]. In the preparation and execution of this operation, the following tasks are allotted to each Service: [...] The task of the Air Force will be:*
  i. *To prevent interference by the enemy Air Force.*
  ii. *To destroy coastal fortresses which might operate against our disembarkation points, to break the first resistance of enemy land forces, and to disperse reserves*

on their way to the front. In carrying out this task, the closest liaison is necessary between individual Air Force units and the Army invasion forces.[72]

Signed: ADOLF HITLER

If Hitler's directive had been successful, or even partly successful, and the Royal Air Force had been destroyed or reduced to a subsidiary role in the conflict, then the following scenarios briefly put forward what may have followed and the possible outcome.

Possible Outcomes
Termination of the RAF as an Effective Fighting Force

With the Royal Air Force defeated and ineffective, British cities and factories and all daily life would have been at the mercy of aerial attack. Invasion would have been very likely, and, if successful, would have led to occupation by Nazi forces and probably the carrying out of similar atrocities to those that had already taken place, and were to take place, in occupied Europe. It would have been the end of Britain as a base and springboard for attacks on German cities and factories, for the supply of war materiel to the USSR through the Arctic convoys and for the D-Day landings in Normandy.

Control of the Mediterranean

Allied operations in and around Italy from 1943 to the end of the war in 1945 could not have taken place. Britain and North Africa would not have been available for use as a base and preparation ground for the invasion of Sicily and Italy. Malta, Gibraltar and Cyprus, indeed the entire Mediterranean, would most likely have fallen under German and Italian control. The Allies' campaign in North Africa could not have taken place because there would have been no forces available to undertake such a venture. The Axis powers of Germany and Italy would have dominated North Africa and the whole of the Mediterranean. All the resources of that region would have been under Nazi control, with the Suez Canal and access to the vast oilfields of the Middle East freely available to the Nazis. With these additional resources, the Axis forces would have become more effective.

U-Boat Dominance in the Atlantic

Great Britain required more than a million tons of imported food and war materiel per week in order to survive and fight during World War II. Had the Nazis invaded and conquered Britain, that million tons of imported goods, mostly from North America, would have stopped. The consequences for the British population would have been cataclysmic.

Churchill was later to state:

The Battle of the Atlantic was the dominating factor all through the war. Never for one moment could we forget that everything happening elsewhere, on land, at sea or in the air, depended ultimately on its outcome.[73]

The shipping losses were enormous during the conflict. Over 72,000 Allied sailors and merchant seamen were killed, and over 3,000 merchant vessels sunk along with

175 warships. The Axis forces lost over 30,000 sailors and almost 800 U-Boats. There would, of course, have been no Battle of the Atlantic had Britain been overrun by the Nazis following defeat at the Battle of Britain. The Royal Navy would, most likely, have been based in Canada, some 3,000 miles from home. What would merchant shipping have done? Certainly not supplied Great Britain with war supplies or food – there would have been no safe landing port available. And with German U-boats able to use British naval bases and be free from air attack in those bases, the Eastern Atlantic, at least, would have become dominated by the Nazis. The World War would have taken on an entirely different dimension.

Effect Upon Morale in Britain
Even without a land invasion from the Nazis, there could only have been a considerable negative reaction. Defeat of the Army at Dunkirk, defeat of the Royal Air Force at the Battle of Britain, imminent prospects of the Wehrmacht crossing the English Channel, continued and uninterrupted bombing of all major infrastructure; the will and determination of the British people would have been stretched.

The British Government, following invasion, would probably have adopted a 'Keep Calm and Carry On' attitude. Indeed, the Government was ready to release 2.5 million 'Keep Calm and Carry On' posters across the country to help minimise possible panic among the population and ensure that there would be no breakdown of essential services. The actual consequences of invasion for the population, however, might well have been very different in Britain than it was for other invaded countries, such as France who, by and large, could feed herself without masses of imported food. One can deduce many things from the prospect of the conquest of the British people. Hunger would have been a certainty for much of the population.

Arctic Convoys of Supply to the USSR
The Arctic convoys of World War II were ocean-going groups of merchant ships with Royal Navy, US Navy and Royal Canadian Navy escorts. They sailed from Britain, Iceland and North America to northern ports in the Soviet Union, primarily Archangel and Murmansk, with millions of tons of much-needed supplies. There were a total of seventy-eight convoys between August 1941 and May 1945 sailing via the Atlantic and Arctic Oceans. The Arctic convoys demonstrated the Allies' commitment to helping the Soviet Union prior to the opening of a Second Front following D-Day. They also tied up a substantial part of Germany's Navy and Air Force in trying to reduce and eliminate the convoys' ability to deliver materiel to the forces of the USSR.

Without these convoys, the USSR would have had much reduced supplies of war materiel. Restocking of arms and food in the USSR would have been delayed or postponed indefinitely. The USSR forces and their allies would have been considerably weaker. They would have been faced with a stronger Axis army because of the additional resources, including oil, available to them following the defeat of Britain, and may well have had to sue for peace with Germany and the Axis powers.

Around 1,400 merchant ships were employed to deliver essential supplies to the Soviet Union under the Lend-Lease programme of aid. The programme was formally known as 'An Act to Further Promote the Defence of the United States' and shipped $11 billion worth of aid to the USSR (approximately $200 billion at 2015 prices). A total of eighty-five merchant vessels and sixteen Royal Navy warships (two cruisers, six destroyers and eight other escort ships) were lost in the operational duration of the Artic convoys. The Nazi *Kriegmarine* lost a number of vessels, including one battleship, three destroyers and at least thirty U-boats, as well as a large number of aircraft.

The Arctic route was the shortest and most direct route for Lend-Lease aid to the USSR, but it was also the most dangerous. Some 3,964,000 tons of goods were shipped by the Arctic route; seven per cent was lost, while ninety-three per cent arrived safely. This constituted some twenty-three per cent of the total aid to the USSR during the war. However, the Arctic route was not the only sea route of supply to the USSR. Other routes were used, including the Persian Corridor and the Pacific Route. Had the Battle of Britain been lost, the Persian Corridor, which was used to supply over 4 million tons of goods to the USSR, could not have been used by the Allies.

The Pacific route opened in August 1941, but was affected by the start of hostilities between Japan and the United States. From December 1941, only Soviet ships could be used, and, as Japan and the USSR observed a strict neutrality towards each other, only non-military goods could be transported. Over 8 million tons of goods went by this route.

Nazi Focus on Defeating the USSR and Linking Up with Japanese Forces in India

This is a difficult scenario for which to forecast a probable outcome. Hitler had suggested in his book, *Mein Kampf*, in 1925 that he would invade the Soviet Union, and Nazism viewed the Soviet Union and most of Eastern Europe as populated by a racially inferior people compared to the Aryan race of Germany. Hitler also declared that the German people needed more living space, i.e. land and raw materials, and that these should be sought in the East. The USSR had one big advantage over the Nazis in that she had overwhelming numbers of personnel available; however many Soviet troops were killed, injured or captured, replacements appeared to be always available. At one stage in the conflict, the Soviet Union was recruiting over 1 million new soldiers every month. Certainly, without the war supplies coming through the Arctic convoys, the war with the USSR would have lasted considerably longer. Had the German onslaught been successful against the Soviets, there would have been little to stop German and Japanese forces meeting on the borders of India and thereby taking control of all of Europe, Asia and North Africa.

Development of an Atomic Bomb

Defeat in the Battle of Britain would have meant that, whatever the ultimate outcome, World War II would have lasted beyond 1945. Hitler would have had much more time to develop other 'Super-weapons of Destruction', including the Atomic Bomb. The Nazis plan to develop a nuclear bomb was not pursued as purposefully as one might imagine. The development programme started in

April 1939, and ended a few months later after many physicists were removed from the programme and drafted into other German initiatives following the declaration of war. A second attempt was begun, but that only lasted until January 1942 when it was assessed that further development could not contribute to the ending of the war in the near term. The programme was not completely scrapped, however, and a further attempt was made under the control of the Reich Research Council. Subsequently, the number of scientists and physicists working on the nuclear programme began to contract, with many having to apply their abilities to more pressing war-time demands.

Opinion between historians is divided on whether Hitler could ever have achieved full development of a nuclear bomb. It could well be argued that it was Hitler's "most obscene and decisive error"[74] to persecute and drive out the Jews from Europe. Many Jewish scientists and physicists, having fled from Nazi persecution during the 1930s, contributed to the manufacture of the first Atomic Bomb. This was developed through the Manhattan Project at Los Alamos in the United States.

With a Germany unfettered by attacks from Britain, the threat of a nuclear Nazi regime would have been much more likely, given that German factories and scientific establishments could have spent more time on developing a nuclear bomb.

To summarise, defeat in the summer of 1940, whether followed by a Nazi invasion of Britain or a peace settlement between Britain and Germany, would probably have resulted in:

- termination of the Royal Air Force as an effective fighting force,
- Axis forces in control of the Mediterranean,
- U-Boat dominance of the Atlantic,
- negative effect upon the morale of Britain,
- no Arctic convoys to supply the Soviet Union,
- increased possibility of defeat of the USSR,
- increased possibility of German development of an atomic bomb.

A recent comment regarding the outcome of the Battle of Britain was:

The Battle of Britain saved the country from invasion. If the RAF had been defeated, all the efforts of the Army and the Navy would hardly have averted defeat in the face of complete German air superiority. With all Europe subjugated, Germany and Japan would later have met on the borders of India.[75]

Alternative Assessments

There are other points of view regarding the importance of the RAF and the Merlin engine, of course, and they must be included here to provide some balance when considering and discussing events from history.

We could, for instance, take what Reichsmarschal Herman Goering – Commander in Chief of the Nazi Luftwaffe – said to one of his interrogators after the end of the war. In response to the question "How did you feel about the outcome to the Battle of Britain?" he replied that it had been an "Indecisive draw".[76] This was because he had been ordered to withdraw his forces from attacking Britain in order to prepare for Operation Barbarossa, the invasion by the Nazis of the Soviet Union.

Well, he would say that, wouldn't he. The Battle of Britain was effectively won when Hitler indefinitely postponed his plan to invade the British Isles, codenamed Operation Sea Lion, on 17 September 1940. It should be noted that Operation Barbarossa did not take place for another nine months, in June 1941.

Not everyone could agree that it was the RAF who were responsible for Hitler's decision not to invade Britain. Another assessment of the situation points out:

Whilst it is true that 'The Few' of the RAF played their part in deterring Germany from seriously contemplating an invasion of Britain, quite how the Germans would have overcome the one true obstacle – the Royal Navy – is unclear. 'The Few' made good copy for newspapers, cinema news scripts and radio broadcasts. After disasters in Norway and France, Britain needed a heroic victory that was easy to understand and dynamic. The Kriegmarine could only muster ten per cent of the Royal Navy's strength, and half its destroyers were destroyed at the port of Narvik in Sweden.

The German high command knew what escort forces it had would be wiped out, and the invasion vessels themselves cut to pieces by the vastly superior Royal Navy. No one can deny the humiliation and defeat inflicted by the RAF on the Germans, but to claim it alone saved Britain from invasion is ludicrous. It is one of the great injustices of recent history that 'The Many' of HMS Rodney *and the Royal Navy's other warships have never been given their due for their part in the Battle of Britain.*[77]

In fact, the *Kriegmarine* lost ten destroyers and other craft during the battle of Narvik in June 1940. In the long summer of 1940, while Britain waited to repel the expected invasion, Winston Churchill, confident in Royal Navy battleships like HMS *Rodney*, with her 16 inch firepower, continued to make himself less than popular with his admirals by insisting that a large naval force should be kept at Rosyth seaport on the Firth of Forth in Scotland. He thought that if the Axis invasion flotillas ever left their French ports, the decisive battle would be fought at sea, not in the air. It would be down to the Royal Navy and its fighting escorts to destroy the invasion boats. Shore leave for the men serving in HMS *Rodney*, and other ships of the Royal Navy, always finished at 6.30pm, because it was essential that the ship knew where her men were every night, and that they were in good physical and mental condition to set sail and attack the German troop transports if the call came. The Royal Navy was, of course, of monumental importance to the eventual outcome of the war, and the ships had to be ever ready.

The war did not end in September 1940. Hitler may have abandoned his immediate plans to invade Britain, though that was not clear at the time. Even the Battle of Britain was not perceived by all at the time as a great victory for the RAF and the country. Group Captain Leonard Cheshire, leader of 617 Squadron of the RAF, believed that invasion was inevitable. In September 1940, he wrote:

> *There's going to be a lot of suffering and death in the next few weeks. I expect a simultaneous invasion of the whole of Great Britain. Let them come.*[78]

The war continued for another five years, and crucial to its eventual outcome was the prowess of the Allies' air forces. Vital to that prowess was the continuing improvement and success in combat of the Allies' aircraft, powered in large part by the Merlin engine.

In the years since 1940 there has been a great deal of revised thinking about the Battle of Britain. The traditional view has always been that British Fighter Command defeated the Luftwaffe against overwhelming odds, and as a result the invasion plans of Hitler were postponed indefinitely. Others query this view, questioning by how much the RAF was outnumbered and emphasising the contribution of the Royal Navy and the threat it posed to a Nazi invasion force.

A no-nonsense view of the multitude of comments and opinions regarding the worth and effectiveness of the battle was given in 2010 by Battle of Britain pilot, Squadron Leader Nigel Rose. When interviewed by the *Herald of Scotland* newspaper on the seventieth anniversary of the battle, he said:

I can understand it. There was exaggeration on both sides, and we didn't see the battle as a turning point towards victory in the war, although it was the first time the Nazis had been halted.

He remembers the threat of invasion as being real, though, and believes the RAF played an important role in preventing it.

We played our part in dissuading the Nazis from launching their invasion. The Navy would question whether we won the Battle of Britain, but we must have had quite an influence because we could still come up and meet the Germans. The Germans reckoned it was going to be a walkover.[79]

Other Factors

While many agree that the contributions of the Hurricane and Spitfire, and therefore of the Merlin engine, were of paramount importance to victory, other factors were of significance to the ultimate outcome. They include:

1. The Germans were fighting a long way from their airbases. This made refuelling and rearming time-consuming. Once engagement with the RAF had begun, it was impossible to do so and get back into the fray as quickly as the RAF pilots did during the battle. Therefore the German fighters had a very limited time that they could spend over Britain before their fuel got too low. The British fighters could land,

refuel and rearm, and be in the air again very quickly, so the RAF had home advantage.

2   The decision of the German High Command to change bombing targets from airfields and RADAR installations (Radio Detection And Ranging) to cities, particularly London. Fighter Command was close to defeat when the switch to attack London and other cities occurred, and so the breathing space this gave was very important. Some accounts suggest that without this change of tactics by the Luftwaffe, the RAF would have been defeated within a very short time.

The single most important factor in the defeat of the Luftwaffe during the Battle of Britain, and subsequently in all other aerial theatres of the war, was the use of the ever improving Merlin engine in the Spitfire and Hurricane fighters of the RAF, and in the numerous aircraft of the Allies' air forces.

The Battle of Britain was the first major campaign to be fought entirely by air forces, and contained the largest and most sustained aerial bombing campaign up to that date. From July 1940, bombers of the German Luftwaffe targeted coastal shipping convoys and shipping ports, such as Portsmouth, as the main targets. In August, the Luftwaffe shifted its attacks to RAF airfields and their supporting infrastructure. As the battle progressed, the Luftwaffe also targeted aircraft factories. The failure of Germany to achieve its objective of destroying Britain's air defences, or of forcing Britain to negotiate an armistice or even an outright surrender, is considered to be the Nazis' first major defeat and probably the turning point in World War II. Germany failed to gain air superiority over the RAF and the threat that Hitler would launch Operation Sea Lion was ended. Great Britain was thereafter to become the foundation from which the Nazi forces were destroyed and the safety and security of the free world was maintained.

As we have seen from comments made at the time, however, the threat of German invasion was very real to those who lived through the battle. With the wealth of opinion about the possible outcomes of the Battle of Britain, from 1940 and on until today, there seems to be general agreement among all historians and commentators. To paraphrase the Duke of Wellington after Waterloo: "It was a very close run thing".[80]

Well, that's almost it. What a story! So many people directly and indirectly involved in the adventure, so many people without whose contribution victory may not have been secured. The parts played by Lord Hives, Sir Stanley Hooker, Arthur Rubbra, Lucy Houston and the many others mentioned herein were vital in this tale of magnificent triumph, a success of an importance that ranks alongside, or even beyond, the turning points in the history of Great Britain secured at Trafalgar, Waterloo and the defeat of the Spanish Armada. The industry and achievement of those individuals helped save the free world.

And today, where are the engineers? Where are the modern equivalents of those great men and women who designed, developed, improved and produced in such abundant numbers the Merlin engine and the aircraft it powered, used to such devastating effect by the very brave men of the RAF and their allies? Where are the future engineers who will walk in the path of Sir Stanley Hooker, Arthur Rubbra, Reginald Mitchell and Tilly Shilling?

Chapter 13
What Now?

Engineering in Britain Today

A large number of graduates in the STEM subjects (Science, Technology, Engineering and Mathematics) is important for Great Britain's future prosperity. Recent figures suggest they are in short supply, however, and there are challenges in recruiting students onto STEM degree courses. The importance of STEM skills to the future prosperity of the country is well established.

Lord Sainsbury's review, titled *The Race to the Top*, in 2007 stated that:

In a world in which the UK's competitive advantage will depend increasingly on innovation and high-value products and services, it is essential that we raise the level of our STEM skills. Policy making in many areas of Government also requires a supply of creative young scientists and engineers.[81]

Another recent report[82] gives the following information regarding students studying engineering in the United Kingdom:

1. Twice the number of engineering graduates is needed to meet the demand for future engineers and engineering teachers.
2. Double the number of young people studying GCSE physics as part of triple sciences and grow the numbers of students studying physics A level to match those studying maths.
3. A big increase in the numbers of pre-nineteen-year-old students studying vocational level 3 qualifications in engineering and manufacturing technology, construction planning and the built environment, and information and communications technologies is required.
4. Engineering companies are projected to have 2.74 million job openings from 2010 – 2020; 1.86 million of these will need engineering skills.
5. The United Kingdom needs to double the number of recruits with Higher National Certificates, foundation degrees, degrees or higher qualifications.
6. The United Kingdom needs to triple the number of apprentices in the sector to 69,000 a year on average; currently 27,000 apprentices a year qualify at level 3.

As for female engineers, their numbers are not quite as low as when Tilly Shilling was at the Royal Aeronautical Establishment in Farnborough, but the latest figures are not encouraging. Why should that be in today's environment? The problem appears to start well before a career is selected. Too many youngsters are not studying maths and science by choice. This trend is especially marked for girls. Almost half of mixed state schools in England recently failed to enter a single girl for A-level physics.[83] The myth that engineering is a dirty and badly paid profession still seems to exist. It might help if schools were to spend some time telling the story of engineering heroes like Hooker, Hives, Mitchell, Shilling, Rubbra and others, and detailing their immense contribution to society and to the world. These engineers are relatively unknown in some parts of Britain, judging by the recent poll[84] taken by the Royal Academy of Engineering. This poll, run by the Royal Academy of Engineering Queen Elizabeth Prize in 2014 to determine who are the heroes of engineering, shows a recognition among the responders of just how wide a topic engineering can be when one looks at the range of engineering disciplines that make the top twenty. Of course a lot depends upon whom one asks to name their engineering hero, and how one asks them. As this poll was conducted using only the social media tool Twitter, one can only surmise how different the list might have been if a wider section of the general public had been involved.

Only one of the people mentioned in this book gets a listing in the poll: Sir Frank Whittle. None of the great engineers who designed, developed and produced the Merlin get into the list. That is quite sad.

The full list of the top twenty is included in Addendum C.

Another recent report highlighted the issue of the shortage of engineers and stated that not enough young people in the United Kingdom are studying STEM subjects to A-level – and as mentioned above, that is especially true of girls. Only twenty per cent of students in England study mathematics beyond the age of sixteen, and only forty per cent of those students are female. The report suggests this is due in part to parental bias in encouraging children towards certain careers, and also "widespread misconceptions and lack of visibility that deter young people".[85]

In addition, not enough engineering graduates are entering into long-term employment in the sector. Three years after graduation, just under seventy per cent of male and fifty per cent of female graduates from engineering and technology programmes are working in their chosen fields. Why this should be so is not clear. It is a fact that only eight per cent of British engineers are women, the lowest figure in Europe. Contrast that figure with Germany: fifteen per cent, Sweden: twenty-five per cent and Latvia: thirty per cent. This is a real problem for British manufacturing, which is currently reduced to just ten per cent of the UK economy.

Danielle George, a Professor of Radio Frequency Engineering at the School of Electrical and Electronic Engineering at Manchester University, has said that young people in Britain today are a generation who can no longer mend gadgets and appliances, because they have grown up in a disposable world and expect everything just to work. According to George, most of today's generation under the age of thirty simply throw faulty appliances away and buy new ones. This is in contrast to previous generations who had to make do and mend. Professor George, who delivered the Royal Institution 2014 Christmas Lectures entitled *Sparks Will Fly: How to Hack your Home*, was only the sixth woman in 189 years to present the Christmas lectures. She followed Susan Greenfield (1994), Nancy

Rothwell (1998), Monica Grady (2003), Sue Hartley (2009) and Alison Woollard (2013). It may be said that six women in the past twenty years is encouraging, though perhaps just not good enough. Future prosperity requires brain-power. More action needs to be taken to improve the situation and encourage extra students into engineering.

We don't know when we might need another Stanley Hooker or Tilly Shilling!

Last Word

The gratitude of every home in our island, in our Empire, and indeed throughout the world, except in the abodes of the guilty, goes out to the British airmen who, undaunted by odds, unwearied in their constant challenge and mortal danger, are turning the tide of the World War by their prowess and by their devotion.

Never in the field of human conflict was so much owed by so many to so few.[86]

Churchill first used these legendary words on 16 August 1940 after visiting the Number 11 Group RAF Operations Room at Uxbridge, near London, during the air battle. Afterwards, travelling back to London by car, Churchill said to his aide, General Hastings 'Pug' Ismay, "Don't speak to me, I have never been so moved." After several minutes of silence, Churchill said, "Never in the history of mankind have so many owed so much to so few."

Ismay replied, "What about Jesus and his disciples?"

"Good old Pug," said Winston, who then changed the

wording to "Never in the field of human conflict..."[87] The famous words would form the basis for Winston's speech to the House of Commons on 20 August 1940. In this speech, Churchill inspired his countrymen by indicating that, although the war had so far been a series of mammoth defeats for Britain and her allies, their situation was now becoming much better.

Although the final words in this heroic tale were to have been Sir Winston's, the following extract from *First Light* by Battle of Britain pilot Geoffrey Wellum DFC sums up much about the Merlin and the contribution of the men and women who helped bring about the survival of Britain at that time, thereby allowing the free world to remain free and develop into that in which we live today, unimpaired by the consequences of defeat to Nazi totalitarianism:

> *I sheer off and follow him round in a turn towards the north-east. Maybe he's had a vector from operations. I look at my watch. We have been about thirty minutes. Just about half time, I reckon, unless Ops recall us, and with a Hun about that's not very likely. The rain is forcing its way through every little gap between the frames and is dripping with the consistency of a Chinese torture over my knees and soaking into my trousers. All the time the Merlin runs without faltering, utterly reliable. What an engine. Thank the Good Lord for Rolls-Royce!*[88]

THE END

Postscript

Merlin (Falco columbarius)
The Merlin is the smallest falcon in Britain. It favours open country such as grassland, dunes and coastal areas, and varies in size between 9 inches and 13 inches in length, with a 20 inch to 29 inch wingspan. Compared with most other small falcons, it is robust and heavily built and punches above its weight. During their nesting period, Merlins are very aggressive towards other birds of prey and towards crows. This is advantageous to the song birds and ground-nesting birds of nearby woods, because during the entire mating season their territory is kept fairly free of flying predators (except for the Merlin, of course). The Merlin is a fierce, energetic predator and is found over much of Britain.[89]

'Robust and heavily built', 'punches above its weight' and 'a fierce, energetic predator': all good definitions of the other Merlin!

Addendum A

Merlin Powered Aircraft

The following is an alphabetical list of all the aircraft that were powered by the Merlin during and immediately after World War II:

- Armstrong Whitworth Whitley, Avro Athena, Avro Lancaster, Avro Lancastrian
- Avro Lincoln, Avro Manchester III, Avro Tudor, Avro York
- Boulton Paul Balliol and Sea Balliol, Boulton Paul Defiant
- Bristol Beaufighter II
- CAC CA-18 Mark 23 Mustang
- Canadair North Star
- CASA 2.111B and D
- Cierva Air Horse
- de Havilland Mosquito, de Havilland Hornet
- Fairey Barracuda, Fairey Battle, Fairey Fulmar, Fairey P.4/34
- Fiat G.59

- Handley Page Halifax, Handley Page Halton
- Hawker Hart (Test bed)
- Hawker Henley
- Hawker Horsley (Test bed)
- Hawker Hotspur
- Hawker Hurricane and Sea Hurricane
- Hispano Aviación HA-1112
- I.Ae. 30 Ñancú
- Miles M.20
- North American Mustang
- Renard R.38
- Short Sturgeon
- Supermarine Type 322
- Supermarine Seafire
- Supermarine Spitfire
- Tsunami Racer
- Vickers F.7/41
- Vickers Wellington Mark II and Mark VI
- Vickers Windsor
- Westland Welkin

This list is from *British Piston Aero Engines and Their Aircraft* by Alec Lumsden. Airlife Publishing, 1994.

Addendum B

Merlin Engine Cutaway Diagram[90]

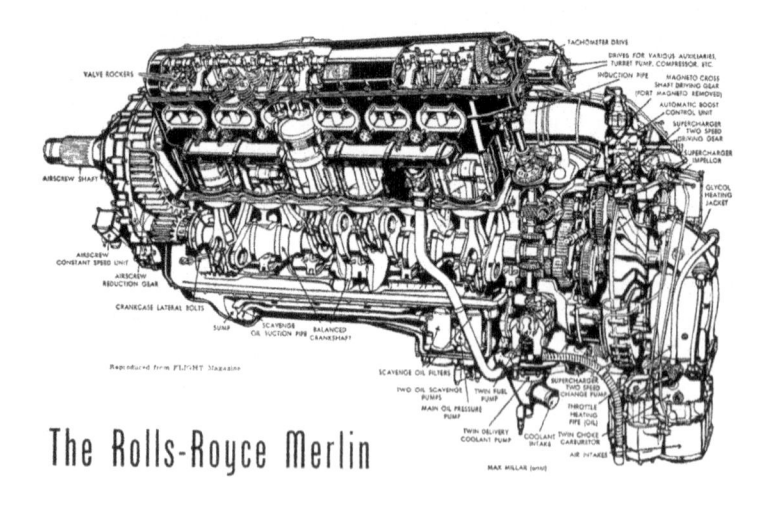

The above cutaway diagram of the Merlin engine is a typical arrangement. It is for basic information only.

Addendum C

Top 20 Engineering Heroes[91]
This list is compiled from the Royal Academy of Engineering feedback from a twitter poll at August 2014. The league table of Top 20 Engineering Heroes was as follows:

1   Isambard Kingdom Brunel (1806–1859), civil and mechanical engineer who built the Great Western Railway and over a hundred bridges, dock systems and ships.
2   Leonardo Da Vinci (1452–1519), Renaissance artist and inventor whose designs included a helicopter, concentrated solar power and flood defences.
3   Nikola Tesla (1856–1943), pioneering electrical engineer who helped develop radio, radar and invented alternating current transmission.
4   Ada Lovelace (1815–1852), mathematician who developed the first computer algorithm.
5   Archimedes (287–212BC), Ancient Greek mathematician, astronomer and inventor.
6   Alan Turing (1912–1954), mathematician and

wartime code breaker who laid out the principles of modern computing and helped develop the first computers.
7   Grace Hopper (1906–1992), computer scientist who designed the first ever English-like data processing language.
8   Alberto Santos-Dumont (1873–1932), aviation pioneer who designed and built the first dirigible and flew the first powered aeroplane in Europe.
9   James Watt (1735–1819), mechanical engineer whose improvements to steam engine technology were fundamental to industrial revolution.
10 Jagadish Chandra Bose (1858–1937), scientist and pioneer of radio communication.
11 Frank Whittle (1907–1966), engineer and air officer who invented the jet engine.
12 George Stephenson (1781–1848), civil and mechanical engineer considered the "Father of Railways" for building the world's first public inter-city railway line to use steam locomotives.
13 Elijah McCoy (1844–1929), one of the first African-American engineers and famous for work on steam-engine lubrication.
14 Joseph Bazalgette (1819–1891), civil engineer who designed the London sewer system.
15 Tim Berners-Lee (1955–present), engineer and computer scientist who invented the World Wide Web.
16 Hertha Ayrton (1854–1923), first female member of the Institution of Electrical Engineers in 1899 famous for research on electric arc.
17 Guglielmo Marconi (1874–1937), electrical

engineer who developed the first long distance telegraph and broadcast the first transatlantic radio signal.

18 Carl Bosch (1874–1940), industrial chemist known for engineering synthetic fertilizer through the Haber-Bosch process.

19 Thomas Edison (1847–1931), a prolific inventor whose most famous devices include the motion picture camera and a long-lasting electric light bulb.

20 Orville (1871–1948) and Wilbur (1867–1912) Wright, aviation pioneers who built and developed the first practical fixed-wing aircraft.

Glossary

Notes on Engineering Terms

It is not intended that this account should be an overtly technical volume. There are plenty of other books[92] available that will explain in detail the effects on an internal combustion aero engine of, for example, multi plate friction clutches or coarse mesh flame traps and the like. This book concentrates more on the main development changes of the Merlin, the engineers and others involved and their impact, and the aircraft and pilots who helped realise the full potential of the Merlin. It may be helpful to readers who are not familiar with the workings of an internal combustion engine, however, if some of the engineering terms used herein are explained.

Carburettor

A carburettor is a device that blends fuel and air for an internal combustion engine. The carburettor relies on suction created by the intake of air accelerated through a *venturi* tube to draw the fuel into the airstream. An engine fitted with a carburettor can deliver a higher specific power output when compared with a fuel injected system due to the lower temperature, hence greater density, of the fuel and air mixture. The mixture is at a lower temperature in a carburetted system because it has not been atomised (pumped through a small nozzle at high temperature) as in a petrol-injected engine, which causes the rise in temperature of the mixture. The amount of air accelerated into the engine can be increased through the use of a supercharger, which increases the power output of the engine.

Superchargers

When piston-engine fighter aircraft fly at higher altitudes, they lose power because the air becomes thinner, and consequently less air is sucked into the engine. This causes a drop in power through decreased combustion. To overcome the reduction in power and allow the fighter to maintain speed at high altitude, a device is needed to suck more air into the engine. Such a device is called a supercharger. It takes the form of a fan, similar to the bellows in a furnace. The power of an engine depends upon the mass of air and fuel it can consume in a given time. A supercharger provides a means of getting more air through an engine, thus providing more oxygen allowing it to burn more fuel, do more work and consequently increase the power of the engine.

Superchargers are mechanically driven by the engine using belts, chains, shafts and gears, or a combination of them, placing a mechanical load on the engine. As an example, in the single-stage single-speed supercharged Merlin, the supercharger used about 150hp. The benefits of using the supercharger outweighed the costs, because the 150hp used to drive the supercharger generated an additional 400hp of power from the engine, giving a net gain of 250hp.

Note: A supercharger is only different from a turbocharger in that it has a different power supply. In a supercharger, power is supplied by mechanical means, usually a belt drive or gears. A turbocharger, however, gets its power from the exhaust stream from the engine. This exhaust stream is directed through a turbine, which in turn spins a compressor. Both supercharger and turbocharger do the same thing, which is to force more air into the engine.

Two-speed Drive Superchargers
Two-speed drives for superchargers allow the engine to keep manifold temperatures low at low altitudes, while at higher altitudes (above 12,000ft), when manifold pressures begin to drop (which reduces the power in the engine), the second speed drive is deployed and the desired manifold pressure achieved. This offers greater flexibility in the operation of the engine, and consequently of the aircraft.

Two-stage Superchargers
In a two-stage supercharger there are two air compressors. The first stage, the low-pressure stage, compresses the air prior to it flowing through a radiator. The radiator cools it before it passes through the second higher pressure stage. Two-stage compressors produce much improved high-altitude performance in the aircraft, allowing greater compression of the air flowing into the engine. Two stage superchargers are almost always two-speed.

Boost

The term 'boost' refers to the amount (in psi – pounds per square inch) by which the intake manifold pressure exceeds atmospheric pressure. This shows the extra air pressure achieved over what would be achieved without the forced induction caused by the use of a supercharger. Atmospheric pressure at sea level is approximately 15lb/f psi. An engine with, for example, 10 psi boost would be forcing air into the engine at 15+10 = 25 psi at sea level. The more air that can be forced into an engine, the better it will perform.

Fuel Injection

Fuel injection is a system for admitting fuel into an internal combustion engine. Fuel injection atomises the fuel by forcibly pumping it through a small nozzle under high pressure before it enters the combustion chamber of the engine. Fuel injection was applied during World War II to almost all aircraft engines made in Germany. These included the widely used BMW801 radial, the inverted inline V12 Daimler-Benz DB601, DB603 and DB605, and the Junkers Jumo 210G, 211 and 213. These engines were used in the Messerschmitt and Focke-Wolf fighters and all Nazi bombers. The major advantages of a fuel injection engine over an engine fitted with a carburettor are increased fuel efficiency and high power output. The gains in power output are achieved by precise control over the amount of fuel through injection timings that are varied according to engine load, thus, in theory, greatly improving efficiency when compared with a conventional carburetted engine. Fuel injection systems were developed by the Bosch engineering company of Germany.

Well, for those of you who are not familiar with the engineering terms used, I hope these few explanations help with your understanding of some of the mechanics involved in the Merlin engine, and indeed all internal combustion aero engines.

Conversion Tables

Imperial	Metric
1 pounds force (lb/f)	4.44822162 newtons
1 pound per square inch (PSI)	6894 n/m² (newton/square metre)
1 yard (36 inches)	0.9144 metres
100 British thermal units (BTUs)	0.0293 kilowatt hours (105500 joules)
1 horse power	745.699872 watts
1 pound (lb)	0.45359237 kilogrammes

Financial Conversion Table

£1 in 1920 = £35 to £265 in 2015
£1 in 1930 = £50 to £350 in 2015
£1 in 1940 = £50 to £220 in 2015

Above figures depend upon whether the comparison is against the cost or value of a commodity, against *income or wealth*, or against a *project*. For more information see: www.measuringworth.com/ukcompare/relativevalue.

Bibliography

Ballantyne, I. *HMS Rodney*: Barnsley, Pen and Sword, 2008.

Banks, R. *I Kept No Diary:* Airlife Publishing Ltd, 1978.

Bingham, V. *Merlin Power – the Growl Behind Air Power in WWII*: Airlife Publishing, 1998.

Brickhill, P. *The Dam Busters*: London, Evans Brothers Ltd, 1951.

Brown, E. *Wings on My Sleeve*: Weidenfeld and Nicholson, London, 2006.

Birch, D. *The Journal of the Rolls-Royce Heritage Trust. Issue 1: The Clitheroe Connection,* May 2014.

Churchill, W. *The Second World War: Triumph and Tragedy: 1943-1945*: Houghton Mifflin 1953.

Dempster, D. & Wood, D. *The Narrow Margin:* Pen & Sword Military Classics (Book 22), 2003.

Evans, C., Whitworth, S., McWilliams, A., and Birch, D. *The Rolls-Royce Meteor*: Derby, Rolls-Royce Heritage Trust, 2004.

Freudenberg, M. *Negative Gravity: A Life of Beatrice Shilling*: Carlton Publications, London, 2003.

Furse, A. *Wilfrid Freeman: The Genius Behind Allied Survival and Air Supremacy, 1939 to 1945*: Staplehurst, Spellmount, 2000.

Gibson, G. *Enemy Coast Ahead,* Introduction by Sir Arthur Harris: Crecy Publishing, 1943.

Goodyear, A. *Something Quite Exceptional*: Derby, Rolls-Royce Heritage Trust, 2010.

Gunston, W. *Rolls-Royce Aero Engines:* Patrick Stephens Ltd, 1989.

Hamilton-Paterson, J. *Empire of the Clouds*: London, Faber and Faber, 2010.

Harvey-Bailey, A. *The Merlin In Perspective – The Combat Years:* Derby, Rolls-Royce Heritage Trust, 1983.

Harvey-Bailey, A. *Hives' Turbulent Barons:* Derby, Rolls-Royce Heritage Trust, 1992.

Hawkes, E. *The Schneider Trophy Contests [1913-1931]:* Southport, R.P. Publications, 1945.

Holter, S. *Leap into Legend: Donald Campbell and the Complete Story of the World Speed Records:* Sigma Leisure Books, London, 2002.

Hooker, S. *Not Much of an Engineer*: Airlife Publishing, 2002.

Isby, D. *The Decisive Duel: Spitfire vs 109:* Little, Brown, 2012.

Johnson, B. *The Churchill Factor*: London, Hodder and Stoughton, 2014.

Kennedy, P. *The Engineers of Victory*: London, Allan Lane, 2013.

Knott, R. *Black Night for Bomber Command*: Pen and Sword, 2007.

Lloyd, Sir I. & Pugh, P. *Hives and the Merlin*: Cambridge, Icon Books, 2004.

Lumsden, A. *British Piston Aero Engines:* London, The Crowood Press, 2005.

Mackay, R. *Lancaster in Action*: Squadron Publications, 1982.

McKinstry, L. *Lancaster: The Second World War's Greatest Bomber:* John Murray, 2009.

McCraw, T. *Creating Modern Capitalism: How Entrepreneurs, Companies and Countries Triumphed in Three Industrial Revolutions*: Harvard Press, 1998.

Mitchell, G. *R.J. Mitchell – Schooldays to Spitfire.* London: Clifford Frost Ltd, 1986.

Mondey, D. *The Hamlyn Concise Guide to British Aircraft of World War II*: London, Hamlyn Publishing, 1982.

Mondey, D. *The Concise Guide to Axis Aircraft of World War II*: London, Chancellor Press, 1996.

Mondey, D. *The Concise Guide to American Aircraft of World War II*: London, Chancellor Press, 1996.

Morris, R. *Cheshire: The Biography of Leonard Cheshire, VC, OM:* London, Viking Books, 2000.

Myhra, D. *Heinkel He 178-Redeaux*: RCW Technology & Ebook Publishing, 28 Sep 2013.

Nockolds, H. *The Magic of a Name:* London, 1972.

Orlebar, A.H. *Schneider Trophy. A Personal Account of High-Speed Flying & the Winning of the Schneider Trophy*: London, 1933.

Pimlott, B. [Editor] *The Second World War Diary of Hugh Dalton:* Jonathan Cape, 1986.

Price, Dr A. *Spitfire: A Documentary History:* TBS,

The Book Service Ltd, 1977.
Pugh, P. *The Magic of a Name – The First 40 Years*: London, Icon Books, 2000.
Roussel, M. *Spitfire's Forgotten Designer:* The History Press, 2013.
Rubbra, A. *Rolls-Royce Piston Aero Engines*: Derby, Rolls-Royce Heritage Trust, 1990.
Rolls-Royce Group PLC. *1904-2004 – A Century of Innovation*: St Ives, Westerham, 2004.
Rolls-Royce Group PLC. *Bristol in Brief*: Bristol, RR Marketing and Information.
Shelton, J. *Schneider Trophy to Spitfire: The Design Career of R.J. Mitchell:* J.H. Haynes & Co Ltd, 2008.
Sherrard, P. *Rolls-Royce Hillington*: Derby, Rolls-Royce Heritage Trust, 2011.
Thompson, K. *H.M.S. Rodney at War*: London, Hollis and Carter, 1946.
Wellum, G. *First Light:* London, Viking Books, 2002.
Cooke, A. *Letter from America – The Letter from Long Island:* BBC Radio, 1970.
Flight magazine, edition of January 13, 1949.
The Derby Telegraph. July 14, 2008. *Reg Spencer comments.*
The Daily Telegraph. June 26, 1999. *Ronnie Harker Obituary.*

Internet Sources

www.en.wikipedia.org/wiki/Never_was_so_much_owed _by_so_many_to_so_few – 2nd Paragraph [Accessed July 2014].

www.en.wikipedia.org/wiki/Keep_Calm_and_Carry_

On#Design_and_production [Accessed July 2014].

www.youtube.com/watch?v=by4lH2whhjk – *Interview with Sir Stanley Hooker* [Accessed July 2014].

www.en.wikipedia.org/wiki/Arthur_Rubbra [Accessed November 2014].

www.en.wikipedia.org/wiki/Arctic_convoys_of_World_War_II [Accessed August 2014].

www.en.wikipedia.org/wiki/Rolls-Royce_Merlin: [Accessed February-April 2014].

www.heraldscotland.com/life-style/real-lives/surviving-the-battle-of-britain-1.1053204 – *An Interview with Nigel Rose* [Accessed November 2014].

www.difp.ie/search – Documents in Irish Foreign Policy No. 196 NAI DFA Secretary's Files A2 [Accessed June 2014].

www.alternatewars.com/WW2/WW2_Documents/Fuhrer_Directives/FD_16.htm [Accessed June 2014].

www.rsbm.royalsocietypublishing.org/content/51/195.full.pdf+html Biographical Memoirs of Fellows of The Royal Society – Lionel Haworth [Accessed November 2014].

www.bonhams.com/auctions/15348/lot/604/ the oldest known surviving Rolls-Royce in the World [Accessed December 2014].

www.telegraph.co.uk/history/world-war-two/10192280/Is-the-Mosquito-the-greatest-warplane-of-all.html [Accessed December 2014].

www.bbc.co.uk/nature/life/Merlin_(bird) [Accessed December 2014].

www.createthefuture.qeprize.org/2014/08/28/engineering hero-countdown [Accessed January 2015].

www.raeng.org.uk/grants-and-prizes/prizes-and-medals/other-awards/queen-elizabeth-prize-for-engineering [Accessed January 2015].

www.motorsportmagazine.com/archive/article/october-1931/33/items-interest [Accessed January 2015].

www.thedambusters.org.uk/gibson.html [Accessed January 2015].

www.geni.com/people/Wing-Commander-Guy-Penrose-Gibson-VC-DSO/6000000014525690633 [Accessed January 2015].

www.allaboutbirds.org/guide/merlin/id [Accessed January 2015].

www.en.wikipedia.org/wiki/List_of_surviving_Supermarine_Spitfires [Accessed January 2015].

www.iop.org/education/teacher/support/girls_physics/file_58196.pdf [Accessed January 2015].

www.enginehistory.org/Piston/Rolls-Royce/RHM/RHM.shtml [Accessed February 2015].

www.flightglobalimages.com/rolls-royce-merlin-xx-cutaway-drawing/print/1569041.html [Accessed March 2015].

Endnotes

Acknowledgements
1   The Rolls-Royce Heritage Trust, PO Box 31, Moor Lane, Derby, DE24 8BJ. Email address: heritage.trust@rolls-royce.com.

Foreword
2   E.W. Hives, General Manager of Rolls-Royce, in a letter to the RR workforce in 1940.

Chapter 1
3   Banks, R. *I Kept No Diary:* Airlife Publishing Ltd, 1978.
4   Shelton, J. *Schneider Trophy to Spitfire: The Design Career of R.J. Mitchell:* J.H. Haynes & Co. Ltd, 2008.
5   Holter, S. *Leap into Legend: Donald Campbell and the Complete Story of the World Speed Records:* Sigma Leisure Books. London 2002.
6   www.rjmitchell-spitfire.co.uk/schneidertrophy/1931.asp?sectionID=2 [Accessed February 2015].

7   Hawkes, E. *The Schneider Trophy Contests [1913-1931]*. Southport: R.P. Publications 1945.
8   www.motorsportmagazine.com/archive/article/october-1931/33/items-interest [Accessed January 2015].
9   Note: while almost all of the sources consulted during research for this book agree that the PV12 engine was the Merlin's direct parent, there are some sources which question this. One of those questioning sources is: *The Rolls-Royce Merlin Aero Engine: Early Development 1933–1937: The Ramp Head Merlin* by J.S.J. Wells, version IV. From www.enginehistory.org/Piston/Rolls-Royce/RHM/RHM.shtml [Accessed February 2015].

Chapter 2

10 Gunston, W. *Rolls-Royce Aero Engines:* Patrick Stephens Ltd, 1989.
11 Harvey-Bailey, A. *The Merlin In Perspective – The Combat Years:* Derby, Rolls-Royce Heritage Trust, 1983.
12 Harvey-Bailey, A. *The Merlin In Perspective – The Combat Years*. Derby, Rolls-Royce Heritage Trust, 1983.
13 www.heraldscotland.com/life-style/real-lives/surviving-the-battle-of-britain-1.1053204 – *An Interview with Nigel Rose* [Accessed November 2014].
14 www.bbc.co.uk/news/uk-england-derbyshire-18781750 – BBC Derby News, 8 August 2012 [Accessed January 2015].

Chapter 3

15 Harvey-Bailey, A. *The Merlin In Perspective – The Combat Years:* Derby, Rolls-Royce Heritage Trust, 1983.

16 Isby, D. *The Decisive Duel: Spitfire vs 109*: Little, Brown, 2012.

Chapter 4

17 Hooker, S. *Not Much of an Engineer*: Airlife Publishing, 2002.

18 Hooker, S. *Not Much of an Engineer*: Airlife Publishing, 2002.

Chapter 5

19 Pugh, P. *The Magic of a Name – The First 40 Years*: London, Icon Books, 2000.

20 Goodyear, A. *Something Quite Exceptional*: Derby, Rolls-Royce Heritage Trust, 2010.

21 McCraw, T, *Creating Modern Capitalism: How Entrepreneurs, Companies and Countries Triumphed in Three Industrial Revolutions*: Harvard Press, 1998, p.113.

22 Furse, A. *Wilfrid Freeman: The Genius Behind Allied Survival and Air Supremacy 1939 to 1945*: Staplehurst. Spellmount, 2000.

23 Furse, A. *Wilfrid Freeman: The Genius Behind Allied Survival and Air Supremacy 1939 to 1945*: Staplehurst. Spellmount, 2000.

24 McKinstry, L. *Lancaster: The Second World War's Greatest Bomber:* John Murray, 2009.

25 Furse, A. *Wilfrid Freeman: The Genius Behind Allied*

Survival and Air Supremacy 1939 to 1945: Spellmount Publishing, 2001.
26 *Time* Magazine *GREAT BRITAIN: Shirts On* Monday, Sept. 16, 1940.

Chapter 6

27 *Oxford Dictionary of National Biography,* Volume 28.
28 *Oxford Dictionary of National Biography,* Volume 28.
29 www.telegraph.co.uk/history/battle-of-britain/8002754/Saviour-of-the-Spitfire.html 2010 [Accessed January 2015].
30 *The Glasgow Herald,* 6 April, 1934.
31 *The Daily Mirror,* Wednesday, 24 October, 1934.
32 *The Daily Telegraph,* 15 September, 2010.
33 www.telegraph.co.uk/history/battle-of-britain/8002754/Saviour-of-the-Spitfire.html [Accessed March 2015].
34 www.bonhams.com/auctions/15348/lot/604/ [Accessed December 2014].
35 http://www.rroc.org/content.asp?contentid=607 [Accessed December 2014].
36 www.team.net/html/fot/1999-12/msg00015.html – *Rolls-Royce Staff During the War*: Accessed March 2015.
37 Rubbra, A. *Rolls-Royce Piston Aero Engines*: Derby, Rolls-Royce Heritage Trust, 1990 – Foreword by L. Haworth.
38 www.youtube.com/watch?v=by4lH2whhjk – *Interview with Sir Stanley Hooker* [Accessed July 2014].
39 www.youtube.com/watch?v=by4lH2whhjk –

Interview with Sir Stanley Hooker [Accessed July 2014].
40 *Lionel Haworth speaking on camera:* www.youtube.com/watch?v=by4lH2whhjk – *Interview with Sir Stanley Hooker* [Accessed November 2014].
41 www.en.wikipedia.org/wiki/Stanley_Hooker [Accessed February 2015].

Chapter 7

42 Price, Dr A. *Spitfire: A Documentary History:* TBS, The Book Service Ltd, 1977.
43 Myhra, D. *Heinkel He 178-Redeaux:* RCW Technology & Ebook Publishing, 2013.
44 Deighton, L. *Fighter: The True Story of the Battle of Britain.* London: Grafton, 1977.
45 Mitchell. G. *R.J. Mitchell – Schooldays to Spitfire*: London, Clifford Frost Ltd, 1986.
46 Roussel, M. *Spitfire's Forgotten Designer:* The History Press, 2013.
47 Knott, R. *Black Night for Bomber Command*: Pen and Sword, 2007.
48 Fowles, C. *NA-73X, The Beginning: the aircraft that changed the course of a war*: The North American P51 Mustang.
49 USA Senate War Investigating Committee report, published 1944.

Chapter 8

50 www.bbc.co.uk/news/uk-england-derbyshire-18781750 [Accessed September 2014].
51 Brown, E. *Wings On My Sleeve*: Weidenfeld and Nicholson, London, 2006.

52 www.telegraph.co.uk/history/world-war-two/1019 2280/Is-the-Mosquito-the-greatest-warplane-of-all. html [Accessed February 2015].
53 www.todayshistorylesson.wordpress.com/tag/ merlin/ [Accessed March 2015].

Chapter 9

54 Orlebar, A.H. *Schneider Trophy: A Personal Account of High-Speed Flying and the Winning of the Schneider Trophy:* Seeley Service Ltd, London, 1933.
55 Gibson, G. *Enemy Coast Ahead.* Introduction by Sir Arthur Harris, 1943.
56 The window was designed by Hugh Easton. Image Copyright Roll-Royce plc.
57 Quote from *Flight* Magazine, January 13, 1949.
58 Goodyear, A. *Something Quite Exceptional*: Derby, Rolls-Royce Heritage Trust, 2010.

Chapter 10

59 Evans, C., Whitworth, S., McWilliams, A., and Birch, D. *The Rolls-Royce Meteor:* Derby, Rolls-Royce Heritage Trust, 2004, p.28.
60 Evans, C., Whitworth, S., McWilliams, A., and Birch, D. *The Rolls-Royce Meteor:* Derby, Rolls-Royce Heritage Trust, 2004, p.28.
61 www.youtube.com/watch?v=by4lH2whhjk – *Interview with Sir Stanley Hooker* [Accessed July 2014]. (Note: for a detailed analysis of the transfer of gas turbine development from the Rover Car Company to Rolls-Royce see *The Clitheroe Connection* by David Birch in *The Journal of the Rolls-Royce*

Heritage Trust, Issue 1, May 2014).
62 www.youtube.com/watch?v=by4lH2whhjk – *Interview with Sir Stanley Hooker* [Accessed July 2014].

Chapter 11

63 Addresses as follows: Retro Track and Air, Upthorpe Lane, Cam, Gloucestershire, GL11 5HP, and Eye Tech Engineering Ltd, 2 Langton Green, Eye, Suffolk, IP23 7HL.
64 For further information please check the BBMF Visitor Centre Website: www.lincolnshire.gov.uk/bbmf.

Chapter 12

65 www.independent.co.uk/voices/comment/biggles-flies-uncensored-more-whisky-less-jingoism-8944480.html [Accessed March 2015].
66 For further information please check the BBMF Visitor Centre Website: www.lincolnshire.gov.uk/bbmf.
67 Queen Elizabeth II, Speech at the Irish State dinner May 2011.
68 Documents in Irish Foreign Policy No. 196 NAI DFA Secretary's Files A2.
69 Pimlott, B. [Editor], *The Second World War Diary of Hugh Dalton: Jonathan Cape, 1986*.
70 *Directive No 16; On the Preparation of a Landing Operation against England*. The full text of this directive is available from: www.alternatewars.com/WW2/WW2_Documents/Fuhrer_Directives/FD_16.htm.
71 Pimlott, B. [Editor], *The Second World War Diary of Hugh Dalton: Jonathan Cape, 1986*.

72 Pimlott, B. [Editor], *The Second World War Diary of Hugh Dalton:* Jonathan Cape, 1986.
73 Churchill, W. *The Second World War: Triumph and Tragedy: 1943-1945:* Houghton Mifflin, 1953.
74 Cooke, A. *The Letter from Long Island:* BBC Radio, 1970.
75 Dempster, D. & Wood, D. *The Narrow Margin:* Pen & Sword Military Classics (Book 22), 2003.
76 Brown, E. *Wings on my Sleeve:* Weidenfeld and Nicholson, London, 2006.
77 Ballantyne, I. *HMS Rodney: Barnsley, Pen and Sword, 2008.*
78 Morris, R. *Letter, 2x7 Sept 1940 LCA BC 51. Cheshire: The Biography of Leonard Cheshire, VC, OM,* London, Viking Books, 2000.
79 www.heraldscotland.com/life-style/real-lives/surviving-the-battle-of-britain-1.1053204 – *Interview with Nigel Rose.*
80 Creevey, T. *Creevey Papers* published in 1903 (What Wellington is reported to have said is: 'It has been a damned nice thing – the nearest run thing you ever saw in your life').

Chapter 13

81 *The Race to the Top: A Review of Government's Science and Innovation Policies* Lord Sainsbury of Turville HMSO, October 2007.
82 Engineering UK. *The State of Engineering 2013.* (Engineering UK is the working name of The Engineering and Technology Board, a company limited by guarantee. Registered in England No.

4322409, Registered Charity No.1089678).
83 www.iop.org/education/teacher/support/girls_physics/file_58196.pdf [Accessed February 2015].
84 www.raeng.org.uk/grants-and-prizes/prizes-and-medals/other-awards/queen-elizabeth-prize-for-engineering [Accessed January 2015].
85 www.topuniversities.com/student-info/daily-news 027/uk-forecasting-shortage-engineers Government report, November 2013 [Accessed October 2014].
86 Hansard. Winston Churchill: House of Commons speech, 20 August, 1940.
87 www.en.wikipedia.org/wiki/Never_was_so_much_owed_by_so_many_to_so_few [Accessed February 2015]. (In 1954 General Ismay related in an anecdote to publisher Rupert Hart-Davis).
88 Wellum, G, *First Light:* London, Viking Books, 2002.

Postscript
89 www.bbc.co.uk/nature/life/Merlin_(bird) [Accessed December 2014] and: www.allaboutbirds.org/guide/merlin/id [Accessed January 2015].

Addendum B
90 www.flightglobalimages.com/rolls-royce-merlin-xx-cutaway-drawing/print/1569041.html [Accessed March 2015].

Addendum C
91 www.raeng.org.uk/grants-and-prizes/prizes-and-medals/other-awards/queen-elizabeth-prize-for-engineering [Accessed January 2015].

Glossary

92 As a good example, try: Harvey-Bailey, A, *The Merlin in Perspective – The Combat Years:* Derby, Rolls-Royce Heritage Trust, 1983.